Statistical Methods for Quality Improvement

Statistical Methods for Quality Improvement

Editor

Jacek Fabian

scitus academics

Statistical Methods for Quality Improvement
Edited by **Jacek Fabian**

ISBN: 978-1-68117-570-6
Library of Congress Control Number: 2016943003

SCITUS Academics LLC,
www.scitusacademics.com
Box No. 4766, 616 Corporate Way,
Suite 2, Valley Cottage,
NY 10989

Printed in United States of America on Acid Free Paper

Preface

Control is very important everywhere. Quality of service provided to customers, management leadership, management commitment, continuous improvement, rapid response, fact-based actions, quality and employee participation are part of quality control. The main objectives of the quality control module are to monitor the reception of materials, claims, customers, internal rejections, evaluations and suppliers of exactly the same corrective actions. These methods and systems guide all quality activities. The use and development of performance indicators is directly or indirectly linked to customer satisfaction and management requirements. Statistical quality control refers to the use of statistical methods to monitor and maintain the quality of products and services.

This book offers many advantages to industry and business as well as a valuable reference for practitioners of statistics, physicist, engineers and who are interested in improving statistical quality.

Table of Contents

Chapter 1

An Integrated Statistical Model to Measure Academic Teaching Quality

Paola Cerchiello, Paolo Giudici

Department of Economics and Management, University of Pavia, Pavia, Italy

ABSTRACT

The aim of this paper is to present a new proposal for the classification of Academic institutions in terms of quality of teaching. Our methodological proposal borrows concepts from operational risk, such as scorecard models, employed to assess University performances, on the basis of both the perceived and the actual quality. We propose to summarize opinion data using new non parametric indexes able to exploit efficiently the ordinal nature of the analysed variables and to integrate different sources of data. In particular we show how web survey methods can improve the quality and robustness of collected data, especially when integrated with students career data. Empirical evidence is given on the basis of real data from the University of Pavia.

Keywords: Teaching Quality; Perceived Quality; Non-Parametric Indexes

1. INTRODUCTION

In this paper we present models for the evaluation of the quality of Academic teaching. The objective of the proposed methodology is to evaluate and, therefore, to improve the quality of University institutions. The quality of University teaching depends mainly on the performances of its students who, thereby, represent the main actors of the analysis. Indeed choosing the right institution could be a key decision for the students future. Moreover University rankings are becoming very important as reputational indicators: an institution with a good ranking position could have a better capacity of attracting students, professors, resources and better chance to participate into important international research projects and meetings. Media attention is very high for rankings such as the Academic Ranking of World Universities (Shanghai Jiao

Tong), as in [1], Times Higher Education World University Rankings, as in [2], D. (2010). Such articles mainly debate on pros and cons of the employed methodologies to produce the relative rankings: the size of the sample size, the typology of survey, the weight assigned to each indicators, the number of indicators. However we believe that also local evaluations should be relevant: they would give the opportunity to an Academic institution to identify, evaluate and classify the risk factors that could affect the global quality of the institution, considering the characteristics and opportunities offered by the territory in which it is located and not only international criteria. To achieve this aim we propose to summarize students perceived quality data (subjective component) and actual quality data (objective component) based on University careers, using new non parametric indexes able to exploit efficiently the ordinal nature of the analysed information. Using such summary indexes, we can obtain a ranking of the taught courses, either considering one aspect per time or in an aggregative way. The final purpose of our analysis is producing indicators that allow the institutional boards, responsible of the organization of teaching activities, to formulate evaluations on the didactic ability of the teachers and consequently planning actions on the organizational component and on the relationship between didactics and adequacy of the resources. The paper is organized as follows: Section 2 describes the context of analysis, Section 3 presents the new proposed indexes, Section 4 shows the application on the available data and Section 5 illustrates the conclusions.

2. WEB BASED QUALITY ASSESSMENT

Assessing the quality of university teaching is a difficult process and the most employed practices analyse student satisfaction questionnaires that measure perceived quality and elaborate them using simple quantitative indicators. Here we present what experimented at the University of Pavia within the context of a research project funded by the Italian evaluation agency CNVSU. In order to improve the teaching quality we have implemented a webbased questionnaire that is compiled by the whole corpus of students in a mandatory way. The experimentation, whose results we are going to present, has been successful: the outcomes have been presented and agreed upon in a National conference and the University of Pavia itself is expected to extended the experimentation to all courses by the and of the year 2013. A relevant innovation of the project is the proposal of new statistical methods to analyse the data that respect their ordinal nature. Before presenting such proposal we inform the reader that after the completion of CNVSU project the authors have entertained further research within the project UNI5NET4STUDENTS granted by the Italian Ministry of Public Administration. Therefore we present the statistical methodologies employed in both the projects and the relative application to the experimental data collected by the University of Pavia.

3. METHODOLOGICAL PROPOSAL

In this section we borrow some concepts from operational risk (e.g. [3-6]). Our aim can be seen as an attempt to present efficient statistical methods aimed at measuring the performance of academic services, through the development of appropriate operational performance and risk indicators. From a managerial aspect, the issue is to properly rank operational risks, say from high priority to low priority, thus to identify fine tuned management decisions and actions directed at improving preventive controls on such risks. In general, the measurement of operational risks leads to the measurement of the efficacy of controls in place: the higher the operational risks, the worse such controls. This points towards a potential efficiency problem. The scorecard approach develops the so-called Self Assessment, based on the experience of a number of internal experts of an institution who usually correspond to different areas of activities or processes. An internal procedure of self assessment can be periodically carried out through questionnaires, submitted to such experts. The collected questionnaires give information on which risks are perceived as most important by the chosen experts, for a future given period. Here we will deal with data derived by questionnaires, compiled by the University students at the end of each course. University students can thus be considered as the experts, evaluating University courses and giving information about the quality of the didactics. Once interviews are collected, the aim is to assign a ranking or rating to each risk event, based on the distribution of the opinions. As described in [7], to assign such ordinal "rating" to the event under analysis (items of the questionnaire), they propose to employ the median as a location measure for each distribution, and the normalized Gini index as an indicator of the "consensus" on such location measure. This results in 4 rating measures for each event, expressed using the conventional risk letters: A for low riskB for medium risk, C for higher risk and D for highest risk. While the median is used to assign a "single letter" measure, the Gini index is used to double or triple the letter, depending on its value.

Several methodologies that take both opinions and data into account, are described for example in [8] and [9]. [8] proposes to quantify and measure operational risk according to a scorecard approach, by collecting expert opinions on each combination business line/event type, in order to overcome the chronic scarcity of quantitative data in such context. [9] propose to measure operational risk by means of a fuzzy approach able to exploit both the qualitative and quantitative data thanks to an interval representation of the data. Recently [10] proposed position indexes based on ordinal variables from discrete parents to efficiently summarize questionnaire items when the number of observations is limited.

In this paper we propose a different approach that exploit two types of information: on one hand quality perception questionnaires and on the other hand the careers of student in terms of velocity in passing exams. The questionnaire is filled by the students in an anonymous way, during the on-line registration to the exam. Since June 2010 it is mandatory to compile the questionnaire at the moment of the exam registration: it is not possible to take an exam without compiling the whole questionnaire. Moreover the questionnaire has two sections: one for the students that actually attended the class under consideration, one for the ones that did not attend. In this way the questionnaire structure permits to evaluate the opinion of the students who really have knowledge on the course contents and the Professor performance. Such questionnaire is considered as a proxy of the quality of the course as perceived by each student, thus representing the subjective component of the analysis being based on personal perceptions. Course quality is to be evaluated not only using perceived quality but also effective quality. According to the standards defined by the Italian Minister of University Education, a University course can be considered of good quality according to different indicators, in particular a rather objective information is the number of years needed to pass the exam. In other words, if students pass the exam related to each course within one academic year, we can conclude that the teaching impact and efficiency is of high level. Conversely, a delay in taking the exam is considered a symptom of low quality. The structure of the two databases, related to the subjective and objective quality, will be described in details in the application section.

Starting from what exposed above, we focus on developing a new method, that allow complete ordering of the University courses exploiting subjective and objective components. A first possible solution is applying a weighted mean of the modalities, as in the following index, named Mean Based Index (MBI):

$$MBI = \frac{N_{dy} + 0.5N_{my} - 0.5N_{mn} - N_{dn}}{N_{dy} + N_{my} + N_{mn} + N_{dn}} \tag{1}$$

where N_{dy}, N_{my}, N_{mn} and N_{dn} point out, respectively, the number of questionnaire responses “definitely yes”, “more yes than no”, “more no than yes” or “definitely no”. The MBI index can then lead to an aggregated value for all items by taking the arithmetic mean among the MBI derived by each question. Note that the MBI index is calculated with default weights for the four possible modalities $(-1,-0.5,+0.5,1)$. Note that the above index can combine the information in the two quality data base by taking, for example, the arithmetic mean of the two corresponding indicators. This because the arithmetic mean is known to possess the associative property. Although the previous index has good properties, our purpose is to obtain new indexes, useful to aggregate different modalities in a single evaluation, but avoiding default weights, that imply an arbitrary quantifications of modalities. Thus we summarize (academic) evaluation data by using an index based on quantiles and frequency excesses of the cumulative distribution function [10]. Such index named Quantile Based Index (QBI) is flexible and depends on the scale employed within questionnaires. QBI is formulated as follows:

$$QBI = \sum_{k=1}^{K} q_k + \left[1 - \frac{\sum_{k=1}^{K-1} \left(F_{(q_k)} - \frac{k}{K} \right)}{50(K-1)} \right] \qquad (2)$$

Or

$$QBI = SC + \left[1 - \frac{TEQ}{\max(TEQ)} \right] \qquad (3)$$

where K is the number of points of the measurement scale, $F(\cdot)$ is the cumulative distribution function,

$\sum_{k=1}^{K} q_k$ (named SC) is the sum of the K points scale that contain the predefined quantiles, TEQ is the sum of the total frequency excesses at each predefined quantile, whose normalization is obtained dividing it by its maximum (equal to 50(K−1)). For example, if we have a 4 points scale for a given question or item, we will employ the 4 associated quartiles (25°, 50°, 75° and 100°) as follows. First we verify in which point scale the first quartile obtains at least 25% of the cumulative distribution function. The obtained point scale is the first term of the SC component. The difference between the cdf calculated in the point scale and the 25% is the first term of the TEQ component. Second we verify in which point scale the second quartile obtains at least 50% of the cdf. The obtained point scale is the second term of the SC component. The difference between the cdf calculated in the point scale and the 50% is the first term of the TEQ component. We proceed analogously for the remaining terms. Note that the part of the index due to the SC quantity establishes the position of the course at hand (our interest unit in this case) on a discrete scale, according to the range reported in **Table 1**. On the other hand, in order to avoid a step function effect, we add the normalized TEQ part that plays the role of a smoother allowing a differentiation among courses sharing the same SC. The proposed index can be normalized according to a [0;1] range. Considering α as the minimum and β as the maximum values assumed by QBI, we can define QBI' as follows:

$$QBI' = \frac{QBI - \alpha}{\beta - \alpha} \tag{4}$$

Note that α ≤ QBI ≤ β and thus $0 \le QBI' \le 1$. For sake of interpretability we can adapt the normalized index to a predefined range, say between γ and δ. For this purpose we set:

$$QBI_n = QBI'(\delta - \gamma) + \gamma \tag{5}$$

From (5), note that

$$\gamma \le QBI_n \le \delta \tag{6}$$

and in our applied context, we set γ = 0, δ = 100 and, therefore,

$$0 \le QBI_n \le +100 \tag{7}$$

For sake of clarity, let us consider the following example: an hypothetic academic course has reported a distribution function for the four modalities as reported in **Table 1**.

On the other hand, we can apply to the data a new non parametric index, built on the basis of the cumulative distribution function alone, which embodies the essential information for ordinal data. We propose to consider the sum of the values of the cumulative distribution function, according to the stochastic dominance approach to model selection that we now describe. It is well know that the cumulative distribution function is a monotonously increasing (or more precisely, non decreasing) function bounded between 0 and 1. On the basis of the cumulative distribution function, a summary index, that we name SDI (Stochastic dominance index) can be calculated as follows:

Table 1. Range of QBI_n index according to different measurement scale.

Measurement Scale	Range (*SC*)	Range (*TEQ*)	Range (*QBI*)
4-points scale	(4;16)	(0;150)	(4;17)
5- points scale	(5;25)	(0;200)	(5;26)
Generalitation	$(K;K^2)$	$(0;(K-1)50)$	$(K;K^2+1)$

$$SDI = \sum_{i=1}^{K} F_i \tag{8}$$

where F is the cumulative distribution function, and K the number of classes. A normalized version SDI' is obtained dividing it by its maximum value, J, which is attained when all data points are concentrated in the lowest class. Note that, this index is based on the cumulative distribution function as the index suggested by Leti.

However, while the SDI index is a second order stochastic dominance measure, based on the arithmetic mean ideal for comparison purposes, Leti's index aims to measure heterogeneity between statistical units. We believe that our proposal is more convenient, because the index maintains the ordinal nature of the data, it is not based on parametric assumptions and its construction is simple to communicate and to interpret.

For communication purposes and comparability to QBI_n we adapted the normalized index to a different range as in Equation (5):

$$0 \leq SDI_n \leq 100 \tag{9}$$

In Tables 2-4 we report an example on how to compute QBI and SDI indexes for a given academic course.

4. APPLICATION

As introduced in Section 3, we consider two different data sets. The first one is related to perceived quality: since February 2010 the first pilot Faculty of the University at hand has adapted a new web-based system to collect the answers of the didactic evaluation questionnaires. At the time of the on-line registration to the exam is given to the student the possibility of compiling the questionnaire regarding that particular course in an anonymous way. Since June 2010 it is mandatory to compile the questionnaire at the moment of the exam registration: it is not possible to take an exam without compiling the whole questionnaire. The questionnaire has two sections: one for the students that actually attended the classes, one for the ones that did not attend. Concerning items, many questions are typically contained in such questionnaire for didactics evaluation. Here we focus our attention on the item that summarizes the overall satisfaction of the student embodied in the following question: E2) "Are you overall satisfied of this course?"

The format of the four-level scale adopted to evaluate courses in our case is the following:

1) Definitely yes (dy);

2) More yes than no (my);

3) More no than yes (mn);

4) Definitely no (dn).

Table 2. Distribution function of a given academic course.

Measurement Scale	Distr. Func	Cdf
Definitely yes	25.47	25.47
More yes than no	47.17	72.64
More no than yes	17.93	90.57
Definitely no	9.43	100.0

Table 3. Sum of the category codes of a given academic course.

25 Quantile	50 Quantile	75 Quantile	100 Quantile	*SC*
1	2	3	4	10

Table 4. Sum of excesses, QBI and SDI.

25%	50%	75%	*TEQ*	*QBI*	*SDI*
0.47	22.64	15.57	36.68	10.74	1.89

We have examined all the on-line data updated to October 2010: the total number of web questionnaires compiled between February and October is 10,929. Before calculating any index we have fixed a threshold to exclude the courses which presented a collected number of questionnaires lower than 5 observations. After this selection and aggregating by course, the data set is composed of 167 observations: each row (statistical unit) represents a course held by a specific Professor.

In **Table 5** we compared subjective and objective quality of a course. The letter A indicates respectively the percentage of "Definitely yes" and the percentage of "Exam taken within less than one year". Thus, we obtain two different ratings regarding a course. Such approach can be useful in a descriptive context and to evidence risk factors, but could be limited in the integration between perceived and actual quality. Thus, we tried to obtain a synthetic score applying the SDI_n and the QBI indexes to both career and web questionnaire results. **Table 6** reports the SDI_n and the QBI_nscores based on the on-line data. From **Table 6** appears that QBI_n re sults to be more severe since only the 9.2% of the courses reaches the highest scores compared to a 13.1% assigned by SDI_nindex. In general QBI_n tends to assign lower scores and it is possible to say that the overall satisfaction expressed by student is good.

The second considered data set collects the data about actual quality: we considered a period of six years, 2004- 2010, and analysed student careers, calculating any delay in taking exams, as previously described. The data available for our analysis contain information about 105,114 passed and not passed exams. As for the perceived quality data set, we aggregated the data to obtain 397 rows, each row representing a specific course. The columns report the percentage correspondent to each value of the four-level scale, for instance "Exam taken within less than one year", "Exam taken within less than two years" and so on. Then we calculated the SDI_n index, in this case about student careers, to obtain a possible ranking in terms of actual quality. Our aim is to integrate both perceived and actual quality, and an evaluation synthetic index. First we matched the two data set, according to the exam identification code. Since our research starts in 2004, during the years not all the courses are held by the same Professor. We chose only the courses with a perfect match between the two data sets: through this selection the courses available for our analysis are 167.

Table 5. Distribution of subjective and objective components according to the 4 ordinal levels.

Label	Subjective	Objective
A	22.7	27.3
B	23.6	29.5
C	30.9	33.3
D	22.7	10.0
TOTAL	100.0	100.0
Median	C	B
Gini Index	0.959	0.853
Final Rating	C	B

Table 6. Distribution of courses based on SDI_n and QBI_n: variable E2.

	SDI_n			QBI_n		
	Freq	Perc	Cum.Perc	Freq	Perc.	Cum.Perc
(20;30]	0	0	0	1	0.60	0.6
(30;40]	1	0.6	0.6	2	1.2	1.8
(40;50]	3	1.8	2.4	6	3.6	5.4
(50;60]	2	1.2	3.6	28	16.8	22.2
(60;70]	14	8.3	11.2	37	22.2	44.3
(70;80]	56	33.3	45.2	37	22.2	66.5
(80;90]	70	41.7	86.9	40	24.0	90.4
(90;100]	22	13.1	100.0	16	9.6	100.0
Total	167	100.0	100.0	167	100.0	100.0

Table 7 reports the SDI_n and the QBI_n scores based on carrers.

As for the questionnaires data, QBI_n offers lower scores, in fact we have 6 courses in the class $[30;50]$, that results to be empty for SDI_n index. Moreover QBI_n scores only 12 courses in the highest class, instead SDI presents 34 courses in the same class. **Table 8** shows such comparison, in terms of quartiles. We recall that the comparison is allowed by the normalization activity applied to the two indexes. The evaluations of the courses drawn through the QBI_n index are more severe in comparison to those of the other indexes. In our opinion, such characteristic of QBI_n index allows a better and more appropriate ordering of the courses that are splined on a wider range of values in comparison to SDI_n. In fact it is evident that SDI_n is less heterogeneous and tends not only to push up courses evaluation but also to produce more optimistic results.

Table 7. Distribution of courses based on SDI_n and QBI_n: based on career.

	SDI_n			QBI_n		
	Freq	Perc	Cum.Perc	Freq	Perc.	Cum.Perc
(30;40]	0	0	0	1	0.6	0.60
(40;50]	0	0	0	5	3.0	3.59
(50;60]	1	0.6	0.6	7	4.2	7.78
(60;70]	6	3.6	4.2	63	37.7	45.51
(70;80]	34	20.4	24.6	51	30.5	76.05
(80;90]	92	55.1	79.6	28	16.8	92.81
(90;100]	34	20.4	100.0	12	7.2	100.0
Total	167	100.0	100.0	167	100.0	100.0

Table 8. Indexes comparison by quartiles.

	SDI_n (Career)	QBI_n(Career)	SDI_n (Quest.)	QBI_n(Quest.)
1 Q	80.32	66.32	75.41	64.96
Media	84.15	74.19	81.00	73.85
3 Q	89.60	76.15	85.71	82.42

Table 9. Rankings based on QBI_n scores sorted by questionnaire.

Course Area	Ranking	QBI_n(Career)	QBI_n(Quest.)
Management	01	100.00	67.05
Management	02	98.21	75.84
Economics	03	98.10	82.18
Economics	04	97.37	66.84
Organizations	05	97.06	91.08
Marketing	163	48.72	74.59
History	164	48.29	73.84
Management	165	38.67	75.61
Marketing	166	30.95	74.83
Informatics	167	23.07	75.06

Table 10. Rankings based on QBI_n scores sorted by career.

Course Area	Ranking	QBI_n(Career)	QBI_n(Quest.)
Economics	01	91.88	82.05
Management	02	91.72	57.21
Management	03	91.57	82.91
Economics	04	91.37	66.21
Management	05	91.33	83.21
Management	163	48.67	57.94
Economics	164	48.67	91.57
Quantitative	165	48.26	65.38
Organizational	166	40.96	57.57
Law	167	39.32	73.84

Table 11. Rankings based on weighted arithmetic mean for SDI_n and QBI_n.

Course Area	Rank	Score SDI_n	Course Area	Rank	Score QBI_n
Law	01	94.19	Organizational	01	93.47
Marketing	02	93.75	Economics	02	89.06
Economics	03	93.62	Law	03	89.26
Economics	04	93.14	Human Res.	04	88.54
Human Res.	05	93.10	Economics	05	87.66
Languages	163	68.00	Economics	163	51.15
Languages	164	66.97	Law	164	49.73
Quantitative	165	63.75	Quantitative	165	48.31
Organizational	166	40.96	Organizational	166	44.56
Law	167	39.32	Law	167	41.28

Thus, we show in Tables 9 and 10 the best and the worse 5 courses ordered respectively according to questionnaires and careers data. In the first positions appears mainly economics and management courses, instead lower positions are more heterogeneous and as expected a quantitative course is present. In particular it is interesting to observe that the worse courses in terms of personal evaluation seem to be not so difficult from the career point of view. Instead the courses more difficult for student received heterogeneous evaluations: in particular one exam of economics has a score equal to 82.18 that is rather high.

We recall that our final aim is to aggregate subjective and objective quality indexes: it is possible to use a weighted arithmetic mean. Since we are considering two data sets, often the number of students considered as a weight for calculating the career is very different in comparison to the number of compiled questionnaires used as the web weight: for instance the course "Marketing" counts for 236 student careers and 132 compiled web questionnaire.

Table 11 presents the courses sorted according to the weighted arithmetic mean (WAM) values. Through this new ranking we notice that appear some courses different from the previous ones. In comparison to the non-aggregated ranking, a good presence of courses regarding "Human resources" and "Organizational" is remarkable among the best courses. On the other hand it is meaningful that among the worst courses, several regards quantitative subjects, in particular mathematics.

5. CONCLUSIONS

In this paper we have presented indexes that can be usefully employed to assess University teaching performances. We started from the operational risk management framework from which we have borrowed the concepts of scorecard approach and ranking. We have compared two possible approaches based respectively on: quantile ranks and frequency excesses on such quantiles and on stochastic dominance. Both the previous methodologies produce ordinal rankings in an effective way. They can be compared with approaches usually employed in University evaluation systems, based on quantitative scorings. Our results show that our ordinal scores are appropriate for comparison and communication purposes. Through the application of such indexes to real data we have obtained a complete ranking of the courses of a pilot Faculty of a medium size Italian University. Future research involves deeper understanding of the mathematical properties of the proposed indexes, especially with reference to extreme values. Another relevant aspect may be the application of what proposed to data concerning reputation as described in [11] in the context of media reputation of corporate companies.

6. ACKNOWLEDGEMENTS

The paper is the result of the close collaboration among the authors. However Paola Cerchiello has written the paper under the supervision of Paolo Giudici. The Authors thank Flavio Ferlini, chief information officer of the University of Pavia, for the data and; Emanuele Dequarti for useful suggestions and the technical support. The Authors acknowledge support from CNVSU project (2009-2010) of the Italian Minister of Research and UNI5NET4STUDENTS project (2010-2011) of the Italian Minister of Public Administration.

REFERENCES

1. N. C. Liu, "The Story of the Academic Rankig of World Universities," International Higher Education, Vol. 54, 2009, pp. 2-3.
2. D. Jobbins, "Global Rankings: Thousands Respond to the Survey," 2010. http://www.universityworldnews.com/article.php?story=20100617192654650
3. M. Cruz, "Modelling, Measuring and Hedging Operational Risk," John Wiley and Sons, Chichester, 2002.

4. C. Alexander, "Regulation, Analysis and Management," Financial Times, Prentice Hall, Upper Saddle River, 2003.
5. J. King, "Operational Risk. Measurement and Modelling," John Wiley and Sons, Chichester, 2001.
6. E. C. Bonafede, P. Cerchiello and P. Giudici, "Statistical models for Business Continuity Management," Journal of Operational Risk, Vol. 2, No. 4, 2007, pp. 79-96.
7. P. Cerchiello, P. Dequarti, P. Giudici and C. Magni, "Scorecard Models to Evaluate Perceived Quality of Academic Teaching," Statistica & Applicazioni, Vol. 2, 2010, pp.145-156.
8. P. Giudici, "Governo dei Rischi: Il Ruolo dei Modelli Statistici," Isituto Lombardo (Rendiconti), Vol. 141, 2007, pp. 361-376.
9. P. Cerchiello and P. Giudici, "Fuzzy Methods for Variable Selection in Operational Risk Management," Journal of Operational Risk, Vol. 7, No. 4, 2012, pp. 1-17.
10. P. Cerchiello and P. Giudici, "On the Distribution of Functionals of Discrete Ordinal Variables," Statistics and Probability Letters, Vol. 82, No. 11, 2012, pp. 2044- 2049.
11. P. Cerchiello, "Statistical Models to Measure Corporate Reputation," Journal of Applied Quantitative Methods, Vol. 6, No. 4, 2011, pp. 58-71.

Chapter 2

Lake Water Monitoring Data Assessment by Multivariate Statistics

Vasil Simeonov[1], Pavlina Simeonova[2], Stefan Tsakovski[3], Vasil Lovchinov[2]

[1]*Chair of Analytical Chemistry, Faculty of Chemistry, University of Sofia "St. Kl. Okhridski", Sofia, Bulgaria*

[2]*Laboratory of Environmental Physics, Georgi Nadjakov Institute of Solid State Physics, Bulgarian Academy of Sciences, Sofia, Bulgaria*

[3]*Chair of Physical Chemistry, Faculty of Chemistry, University of Sofia "St. Kl. Okhridski", Sofia, Bulgaria*

ABSTRACT

The application of multivariate statistical methods to high mountain lakes monitoring data has offered some important conclusions about the importance of environmetric approaches in lake water quality assessment. Various methods like cluster analysis and principal components analysis were used for classification and projection of the data set from a big number of lakes from Pirin Mountain in Bulgaria. Additionally, self-organizing maps of Kohonen were constructed in order to solve some classification tasks. An effort was made to relate the maps with the input data in order to detect classification patterns in the data set. Thus, discrimination chemical parameters for each pattern (cluster) identified was found, which enables better interpretation of the ecological state of the system. A methodology for application of combination of different environmetric methods was suggested as a pathway to interpret high mountain lake waters monitoring data.

1. INTRODUCTION

The careful monitoring of natural water systems like river streams, lakes, wells, and underground sources is a very responsible task. Usually, a set of chemical and physicochemical parameters reflecting the surface or underground water quality are carefully analysed and the results obtained are compared to certain threshold values in order to decide if the water quality meets the quality desired. The choice of quality parameters is normally standardized and described in various instructions and directives for individual countries or unions like EU [1-4]. Recently, a very specific attention requires a different type of water quality parameter called “toxicity” (“ecotoxicity”) [5-7]. It is also important to note that lake waters are important element of the aquatic ecosystems. They constitute ecological niches (especially in combination with the lake bottom sediments) supporting not only fish but also benthic organisms, i.e. animals and plants living on the bottom of bodies of water, and are a source of nutrients for aquatic organisms such as small invertebrates and protozoans. An assessment of the effect of pollution on life in lake water bodies requires also monitoring of bottom sediment samples both for chemical, physicochemical, and toxicity parameters. They are very useful material for various environmental studies because they act as sorption column and provide a clear image of all events taking place in the overlying water layer.

Very often, however, the monitoring data are considered in a “univariate” way—each parameter separately. In the reality the state of an ecosystem is depending simultaneously on many factors and parameters. Therefore, these systems are multivariate in nature. That is why the classification, modeling and interpretation of the monitoring data sets have to be performed by the use of the chemometrics and environmetrics [8-14], where the references given are only a tiny part of many environmetric studies. The specific point in the studies of lake waters is that there is lack of intelligent data analysis of the monitoring sets comprising of different water quality parameters simultaneously interpreted.

The aim of the present chapter is to demonstrate the role of environmetric classification, modeling, and interpretation of monitoring data from the lake systems of Pirin Mountain, Bulgaria.

2. CHEMOMETRIC METHODS

The modern chemometrics is a branch of chemistry (very often related to analytical chemistry) which deals with the application of mathematical and statistical methods in order to evaluate, classify, model, and interpret chemical and analytical data, to optimize and model chemical and analytical processes and experiments and to extract a maximum of chemical and analytical information from experimental data. When the methods of chemometrics are applied to data sets obtained by monitoring of various environmental compartments (surface water, atmosphere, soil, sediments, biota etc) the term environmetrics is used to stress the information ability of the methods to gain specific information from samples of the total environment.

Very important methods of multivariate statistics employed in environmetrics are cluster analysis, principal components analysis and self-organizing maps of Kohonen which will be briefly presented and applied.

Cluster analysis (CA) is an exploratory data analysis tool for solving classification problems [15]. CA enables objects stepwise aggregation according to the similarity of their features. As a result hierarchically or non-hierarchically ordered clusters are formed. Primary standardization of features becomes necessary to avoid effects of dimensionality on the classification results. There are a variety of different measures of inter-cases distances and inter-cluster similarities and distances to use as criteria when merging nearest clusters into broader groups or when considering the relation of an object to a cluster. The most applied is the Euclidean distance (after the standardization of the raw data).

In case of CA one task is related with determination of similarity between measured objects as well as similarities between the features describing the objects. As in case of distance measure various algorithms (linkage techniques) are available to decide on the number of clusters. Very popular linkage algorithm in hierarchical clustering problems is Ward's method [15].

In hierarchical agglomerative clustering the graphical output of the analysis is usually a dendrogram—a tree-like graphics, which indicates the linkage between the clustered objects with respect to their similarity (distance measure). For practical reasons the Sneath index of cluster significance is widely used. It represents this significance on two levels of distance measure D/D_{max} relation: $1/3D_{max}$ and $2/3D_{max}$. Only clusters remaining compact after breaking the linkage at these two distances are considered significant and are object of interpretation.

Principal Component Analysis (PCA) seems to be the most widespread multivariate chemometric technique and is a typical display method (also known as eigenvector analysis, eigenvector decomposition or Karhunen-Loéve expansion). It enables revealing the "hidden" structure of the data set and helps to explain the influence of latent factors on the data distribution [16,17]. PCA transforms the original data matrix into a product of two matrices, one of which contains the information about the objects and the other about the features. The matrix characterizing objects contains the scores (understood as projection) of objects on principal components (PCs). The other one, characterizing features is a square matrix and contains the set of eigenvectors (understood as weights, in PCA terminology called "loadings") of the original features in each PC.

Some important features of PCA could be summarized as follows. The principal components axes (the axes of the hidden variables) are orthogonal to each other. Most of the variance of the data is contained in the first principal component. In the second component there is more information than in the third one etc. For interpretation of the projected data both the score and the loading vectors are plotted. In the score plots, the grouping of objects can be recognized. A loading plot reveals the importance of the individual variables with respect to the principal component model.

A very important task in PCA is the estimating the number of principal components necessary for a particular PC model. Several criteria exist in determining the number of components in the PCA model: percentage of explained variance, eigenvalue—one criterion, Scree— test, cross validation [16,17].

Interpretation of the results of PCA is usually carried out by visualization of the component scores and loadings.

Self Organizing Map (SOM) algorithm has been proposed by Kohonen in 1980 [18]. It is a neural-network based model which shares, with the conventional ordination methods, the basic idea of displaying a high-dimensional signal manifold onto a much lower dimensional network in an orderly fashion (usually 2D space). The most common shape of the Kohonen map is a rectangular grid with the number of hexagonal nodes.

When a plenty of features is considered it is difficult to compare all maps for all features and thus becomes necessary to find similarity between them, and simultaneously, in the cases' space and classify them into clusters. Input features' planes (e.g. variables) could be visualized on a summary SOM map (called also as unified distance matrix or U-matrix) to show the contribution of each feature in the self-organization of the map. U-matrix visualizes distances between neighboring map units, and helps to identify cluster structure of the map. The U-matrix joined with features' planes can be effectively applied for assessment of inter-features and inter-cases relations. Finally, the best classification with the lowest Davies-Bouldwin index should be chosen (it is a function of the ratio of the sum of within-cluster scatter and between-cluster separation).

Main advantages of SOM algorithm application are: Semi-quantitative information about the distribution of a given feature in the space of the cases; visualization of similarity between positive as well as negative correlated features; visualization and classification of "outliers" i.e. those features or cases which do not belong to a wellorganized, homogeneous populations; SOM is noise tolerant (this property is highly desirable when site-measured data are used).

3. RESULTS AND DISCUSSION

The application of the multivariate statistical methods described above as one of the most important tool in assessment of lake water quality can be illustrated by specific case studies. Thus, the classification, data projection, modeling and interpretation of lake water monitoring data sets becomes understandable and turns to be a pattern to follow in lake pollution research.

The data was collected during expeditions in 2001 and 2002 for big number of lakes located in Pirin Mountain, which is one of the highest mountains in Bulgaria. The sampling sites and their heights are indicated in **Table 1**. The sampling period was between May and October. The sampling itself was performed on the lake surface approximately 2 m from the costal line. The water samples (about 100 mL) were placed in polyethylene flasks. The chemical analysis was carried out within 4 days after sampling at Faculty of Chemistry, University of Sofia. Altogether eleven chemical parameters (major cations and anions like sodium, potassium, calcium, magnesium, chloride, sulfate, nitrate, hydrogen carbonate) were analysed by electrothermal atomic absorption spectrometry and ion chromatography as well as pH (potentiometrically), conductivity (conductometrically), water temperature, and dissolved matter (by summing up of chemical concentrations). Due to the rapid changes of some parameters like pH, temperature, conductivity their determination was done directly at the sampling site by the use of portable instruments. The number of lakes involved in this study was over forty. It is worth to mention that water samples were taken not only from the lakes but also from rivers and springs in the vicinity of the lakes in order to obtain a more realistic estimation of the water quality and of the various natural and anthropogenic impacts.

The data sets from the lakes from the two mountains were classified, modeled and interpreted by the use of cluster analysis, principal components analysis and self-organizing maps of Kohonen. The goal of the environmetric interpretation was to identify groups of similarity between the lakes, to find relationship between the chemical parameters for the lake water quality, to detect hidden factors responsible for the data structure as well as to reveal discriminating chemical parameters, which determine the separation of the lakes in different groups of similarity (or dissimilarity).

The initial data set for the samples from Pirin lakes includes 74 objects (sampling locations and sampling periods). No normal distribution of data (for each of the variables) is found, which means that the input data were subject of standardization before applying multivariate statistics (**Table 2**).

The first step in the classification of the monitoring data was clustering of the sampling sites (Ward's method of linkage, squared Euclidean distance as similarity measure, z-transformation of data and check of the cluster significance by the Sneath index). The hierarchical dendrogram is shown in **Figure 1**.

It is readily seen that a good homogeneity of the water quality for all lakes is found. There is a tiny group of outliers (for the more strict Sneath's criterion of $2/3D_{max}$) mostly from the region of Lake Sinanitsa and Sinanitsa River characterized by quite specific lithoral composition.

In **Figure 2** the hierarchical dendrogram for linkage of water quality parameters is presented. Four clusters are formed as follows: (nitrate, chloride, sulfate), (potassium, sodium), (calcium, magnesium, conductivity), (hydrogen carbonate, dissolved matter) and pH as an outlier. It could be concluded from the cluster analysis results that factors like water hardness, water acidity, salt content and turbidity are responsible for the relative separation of the sampling events. However, this information is not convincing enough to make final decisions on the lake water quality.

Table 1. Short description of the lakes in Pirin subject to assessment.

Code	Sampling site	Height a.s.l.[m]
P1	Lake Suhodolsko	2311
P3	Lake Dolno Todorino	2510
P4	Lake Gorno Todorino	2536
P5	Lake Dalgo Banrerishko	2310
P6	Lake Djabeshko Banderishko	2322
P7	Lake Banderishko	2312
P8	Lake Muratovo	2230
P9	Lake Spanopolsko	2302
P10	Lake Dolno Georgiisko	2304
P11	LakeGorno Georgiisko	2392
P12	Sinanitsa River	
P13	Lake Sinanitsa	2181
P14	Sarchaliiska River	
P15	Lake Chairsko	2355
P16	Lake Prevalsko	2305
P17	Lake Prevalsko	2312
P18	Lake Tevno Belmetsko	2512
P19	Lake Samnodivsko 2	2375
P20	Snow from Samodivski circus	
P21	Lake Samnodivsko 3	2372

P22	Lake Popovo	2234
P23	Lake Kremensko	2356
P24	Lake Gorno Kremensko	2352
P25	Lake Dolno Kremensko	2304
P26, P27	Lake Bezbog	2239
P28	Lake Popovo 6	2185
P29	Lake Popovo 3	2208
P30	Lake Valiavishko	2419
P31	Lake Goliamo Valiavishko	2280
P32	Lake Dolno Valiavishko	2254
P33	Spring Moiseeva cheshma	2370
P34	Lake Gorno Vasilashko	2154
P35	Lake Ribno Vasilashko	2162
P36	Lake Dolno Vasilashko	2325
P37	Lake Gorno Tipitsko	2445
P38	Lake Bashliisko	2450
P39	Lake Bashliisko 6	2430
P40	Lake Bashliisko 3	2461
P41	Lake Bashliisko 4	2313
P42	Lake Begovishko	2392

Table 2. Basic statistics for all Pirin lakes.

Variable	N	mean	min	max	S.D.	skewness
pH	74	6.7	0.50	17.0	3.0	0.8
DM	74	10.0	0.61	50.9	7.9	2.4
Conductivity	74	18.9	3.00	108.3	19.1	2.8
Ca^{2+}	74	28.8	6.00	234.0	44.3	3.7
Mg^{2+}	74	8.1	2.60	45.6	8.3	3.5
Na^{+}	74	23.2	5.60	65.9	11.4	1.2
K^{+}	74	2.2	0.00	16.0	2.2	3.7
HCO_3^-	74	76.0	0.80	623.6	95.0	3.0
SO_4^{2-}	74	23.7	1.90	104.1	17.6	1.3
Cl^-	74	7.1	0.30	39.7	7.4	1.8
NO_3^-	74	1.8	0.01	16.1	3.5	2.4

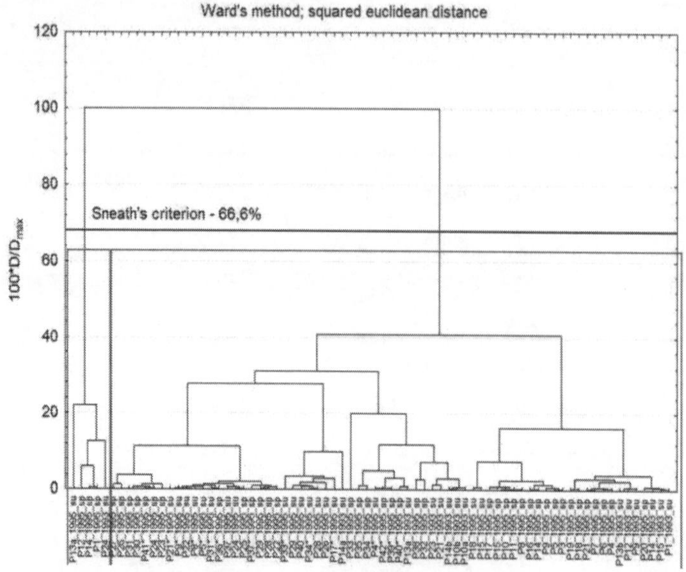

Figure 1. Hierarchical dendrogram for sampling locations.

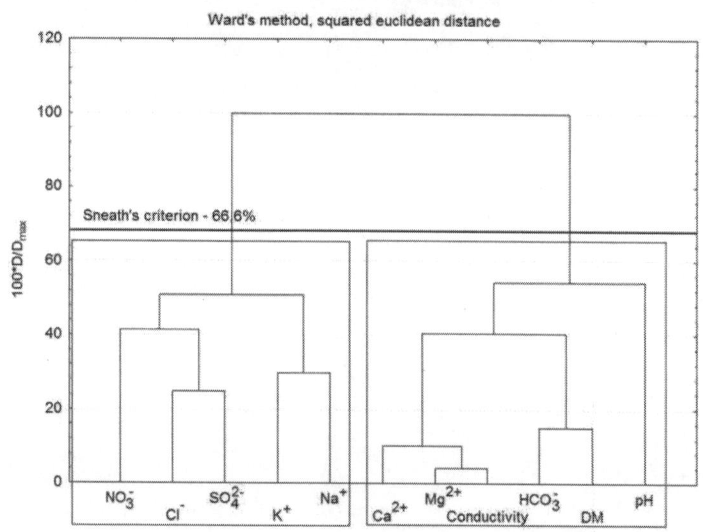

Figure 2. Hierarchical dendrogram for variables.

In order to gain additional information from the data set PCA was performed on the normalized input data. Firstly, the factor scores plot will be considered (**Figure 3**).

The homogeneity of the data (very close levels of most of the chemical parameters characterizing the lake water quality) is indicated again by the big cloud of similar objects. The outliers marked are from the specific dolomite circus where the lakes Sinanitsa, Suhodolsko and Gorno Kremensko are located. Again, the separation is due to a geographical (natural) rather than anthropogenic impact.

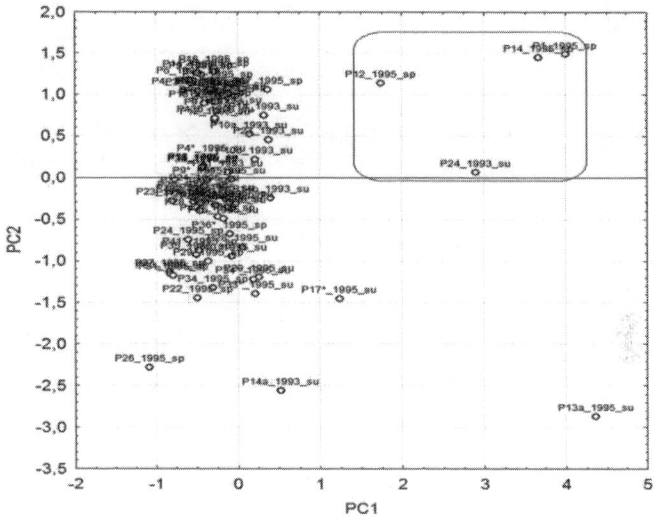

Figure 3. Factor scores plot (PC1 vs. PC 2).

Figure 4 (PC1 vs. PC2) for the factor loadings (normalized data, Varimax rotation mode) indicates the grouping of the chemical variables with respect to the identified latent factors. It is seen that high correlation along PC1 axis is found for chloride, sulfate and hydrogen carbonate and along PC2 axis—for calcium, magnesium, conductivity and dissolved matter. These results slightly contradict those found by the classification with cluster analysis and that is why a more careful inspection of the PCA results is needed.

In **Table 3** the factor loadings for the identified four latent factors are presented. These four factors explain over 75% of the total variance of the system. The first one could be conditionally named "water hardness" factor since it indicated high correlation between calcium, magnesium, conductivity, and dissolved matter. The second latent factor is related to hydrogen carbonate, sulfate and chloride and, logically, its conditional name might be "anthropogenic impact" since it indicates possible pollution by atmospheric transfer of secondary emissions. The third factor should be attributed to processes of lake water acidification by the negative correlation between pH and nitrate.

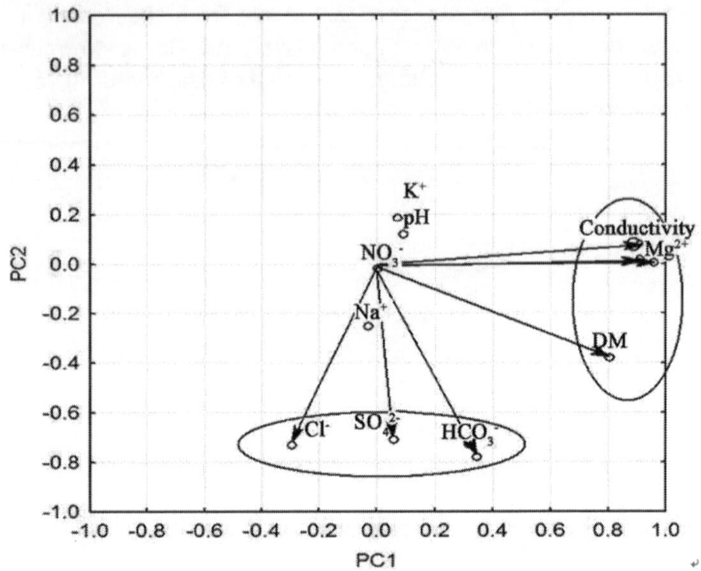

Figure 4. Factor loadings plot (PC1 vs. PC 2).

Table 3. Factor loadings(normalized Varimax Rotation).

Variable	PC1	PC2	PC3	PC4
pH	0.093	0.119	**0.737**	-0.160
DM	**0.805**	-0.378	0.292	0.079
Conductivity	**0.915**	0.078	-0.161	0.117
Ca^{2+}	**0.909**	0.017	0.122	-0.039
Mg^{2+}	**0.961**	0.003	-0.002	-0.037
Na^{+}	-0.030	-0.252	0.139	**0.820**
K^{+}	0.072	0.184	-0.212	**0.756**
HCO_3^-	0.348	**-0.778**	0.100	0.050
SO_4^{2-}	0.059	**-0.707**	-0.313	0.385
Cl^-	-0.295	**-0.730**	-0.115	-0.199
NO_3^-	0.004	-0.018	**-0.818**	-0.076
Variance explained %	31.5	17.3	13.9	13.5

The fourth and the last latent factor is characterized by high (statistically significant loadings) for sodium and potassium and could be conditionally named "salt" factor.

The separation of several outliers by CA and PCA and the identification of the latent factors responsible for the data structure could be improved and completed by data classification using SOM. In **Figure 5** the self-organizing maps for all variables and all sampling events are shown. Similarities detected by Ca and PCA are also obvious on some of the maps—e.g. conductivity, calcium, magnesium, and dissolved matter maps reveal one and the same pattern—the highest concentration levels are grouped in the upper part of the maps.

In **Figure 6** the SOM grouping of the variables is presented. The grouping is slightly different as compared to the PCA results or CA classification. Very stable (the same for all classification approaches) is the group consisting of calcium, magnesium, dissolved matter, and conductivity. This is an important indication for the role of this relationship for the lake water quality assessment of Pirin region. In **Figure 6** one could detect similarity between hydrogen carbonate and sodium (not with sulfate and chloride), which is rather an indication for lithoral impact than acidification one. Thus, the different types of classification of the monitoring data indicate possibly different pathways in correlation between the water quality parameters and, parallel to it, the stable for all classification patterns.

Very important step is to divide the data set into significant clusters with their spatial vicinity (possible only by SOM classification) and, further, to detect the discriminate parameters for each one of the clusters formed. In **Figure 7** the optimal number of clusters and the respective hit diagram are shown.

For all monitoring data from Pirin lakes three clusters are formed. Their content can be easily determined along with the discriminating tracers for each one of them (**Figure 8**).

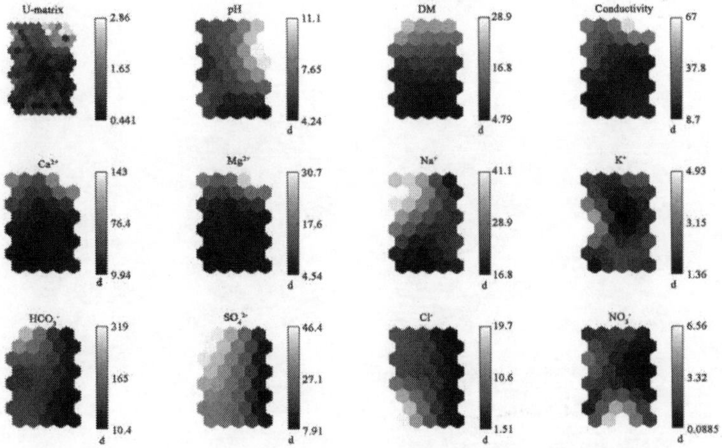

Figure 5. SOM classification for each variable at all sampling locations.

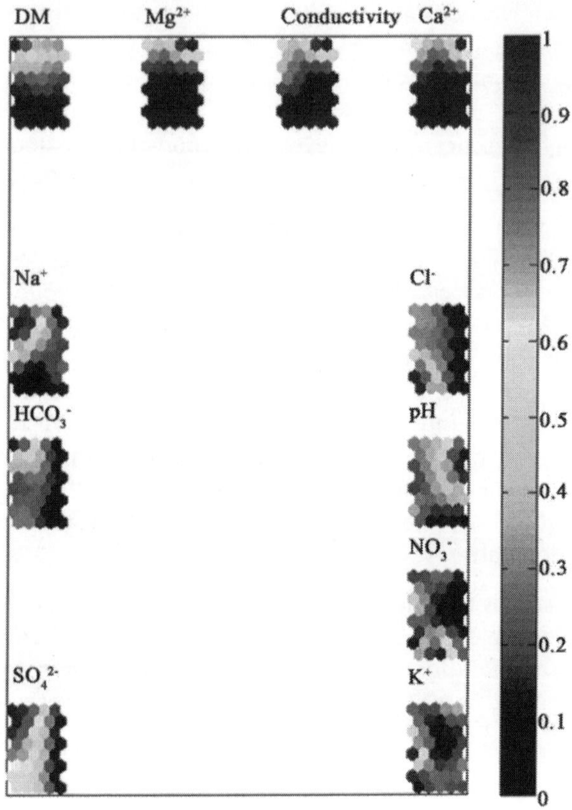

Figure 6. SOM of grouping of variables.

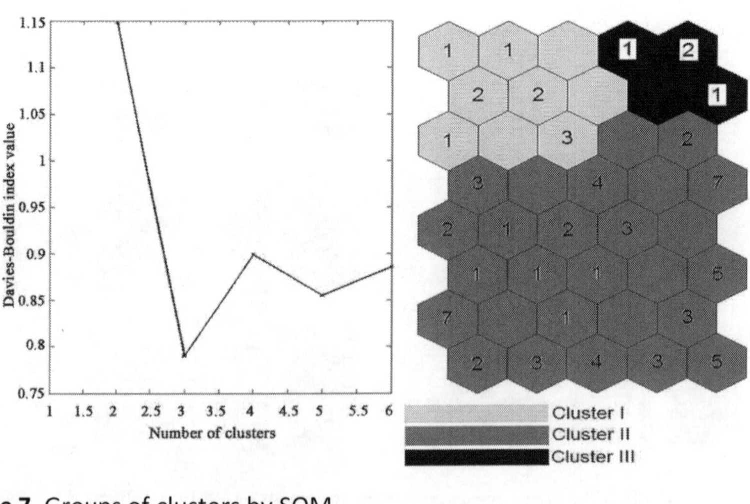

Figure 7. Groups of clusters by SOM.

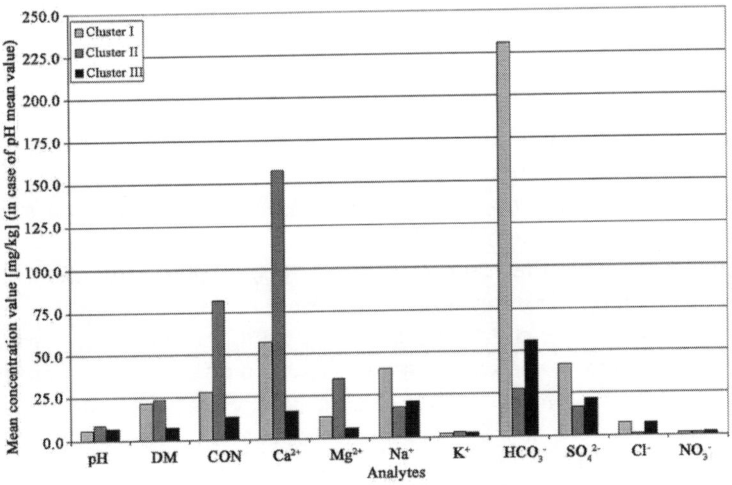

Figure 8. Discrimination diagram for all 6 SOM clusters.

Cluster I includes dominantly lakes from the southern part of Pirin Mountain (the group of Vasilashki lakes, which is located at relatively lower levels a.s.l. as compared to the other lakes). All objects are samples from the summer season (lowest water level) and are characterized by highest concentrations of hydrogen carbonate, sulfate, sodium, and chloride. This is an indication for high salt impact but it should be attributed to natural and not to anthropogenic reasons.

Cluster II comprises a small group of lakes from the Lake Sinanitsa and Sinanitsa River vicinity. They are the outliers discussed by the interpretation of results from CA and PCA of the monitoring data. The discriminating variables in this situation are the parameters of the water hardness – calcium, magnesium, conductivity, and dissolved matter. Obviously, the specific local crustal composition is the natural reason for finding similarity between this group of Pirin lakes.

Cluster III is mixed and very large. This is an indication for the homogeneity of the water quality stated before. The group of more than 60 cases is characterized by lowest levels of the concentrations of all water quality parameters.

In conclusion, it can be stated that the quality of the lake water in the high-mountain lakes in southern Bulgaria is very high and is mostly subject to natural impacts. The different environmetric data interpretations applied prove convincingly this statement.

4. CONCLUSIONS

In this study we have tried to prove that the lake water monitoring data assessment could use an extremely useful tool—the methods of chemometrics and environmetrics. If one decides to use the multivariate statistical methods for classification, projection, modeling, and interpretation of lake water monitoring data in order to gain specific (often hidden, non-available from the raw data) information, then the pathway to reach it should follow several major steps: proving the data quality by means of metrological criteria (uncertainty, limits of detection of the monitoring methods, precision, reliability of signal etc.); checking the data distribution (often non-normal) by statistical tests; data normalization in order to avoid problems with non-normal distribution or data dimensionality; classification of the monitoring data by different environmetric approaches like cluster analysis, principal components analysis, neuron net classification (self-organizing maps of Kohonen as an option of classification without training procedures); careful interpretation of the classification results and finding the reasons for the similarity groups; modeling the data in order to identify latent factors responsible for the data structure and the factor contribution to the formation of the total concentration of each of the lake water quality parameters; determination of seasonal patterns in lake water quality; comparison of the models with the real monitoring data.

Following this mode of data interpretation we have reached to the important conclusion that reliable classification for high-mountain lakes in Pirin Mountain is performed, which makes it possible to find patterns of similarity between the lakes and to explain this similarity (or dissimilarity) by discriminating water quality parameters; it could be stated that most of the Pirin high-mountain lakes are clean with background levels of chemicals.

5. ACKNOWLEDGEMENTS

The authors would like to express their sincere gratitude to the National Science Fund (Project DO-02-352) for the financial support which made this study possible.

REFERENCES

1. EU Instruction 80/778, “Water Analysis and Control,” Brussels, 1990.
2. Council Directive 91/692/EEC, European Union Directive, 1991.
3. Bulgarian Drinking Water Analysis Standard (BDWS), “Ministry of Environment and Water,” Sofia, 1991.

4. European Commission Directive 2000/60/EC of the European Parliament and of the Council, "Community Action in the Field of Water Policy," Official Journal of the European Union, Serie L327, 2000.

5. V. Simeonov, L. Wolska, A. Kuczynska, J. Gurwin, S. Tsakovski, M. Protasowicki and J. Namiesnik, "Sediment-quality Assessment by Intelligent Data Analysis," Trends in Analytical Chemistry, Vol. 26, No. 4, April 2007, pp. 323-331.

6. US Environmental Protection Agency, "Clarification Regarding Toxicity Reduction and Identification Evaluations in the National Pollution Discharge Elimination System Program," Washington, D.C., 2001.

7. G. Goodfellow, L. Ausley, D. Burton, D. Denton, P. Dorn, D. Grothe, M. Heber, T. Norber-King and J. Rogers, "Major Ion Toxicity in Effluents: A Review with Permitting Recommendations," Environmental Toxicology and Chemistry, Vol. 19, No. 1, January 2000, pp. 175-182.

8. A. Astel, S. Tsakovski, P. Barbieri and V. Simeonov, "Comparison of Self-organizing Maps Classification Approach with Cluster and Principal Components Analysis for Large Environmental Data Sets", Water Research, Vol. 41, No. 19, November 2007, pp. 4566-4578.

9. V. Simeonov, L. Wolska, A. Kuczynska, J. Gurwin, S. Tsakovski and J. Namiesnik, "Chemometric Estimation of Natural Water Samples Using Toxicity Tests and Physicochemical Parameters," Critical Reviews in Analytical Chemistry, Vol. 37, January 2007, pp. 81-90.

10. A. Astel, S. Tsakovski, V. Simeonov, E. Reisenhofer, S. Piselli and P. Barbieri, "Multivariate Classification and Modeling in Surface Water Pollution Estimation," Analytical Bioanalytical Chemistry, Vol. 390, No. 5, March 2008, pp. 1283-1292.

11. P. Simeonova and V. Simeonov, "Chemometrics to Evaluate the Quality of Water Sources for Human Consumption," Microchimica Acta, Vol. 156, No. 3-4, December 2006, pp. 315-320.

12. A. Astel, G. Glosinska, T. Sobczynski, L. Boszke, V. Simeonov and J. Siepak, "Chemometrics in the Assessment in the Sustainable Development Rule Implementation," Central European Journal of Chemistry, Vol. 4, No. 3, September 2006, pp. 543-564.

13. P. Simeonova, "Polluting Sources Apportionment for Atmospheric and Coastal Sediments Environment," Ecological Chemistry Engineering, Vol. 13, 2006, pp. 1021- 1032.

14. P. Simeonova, "Multivariate Statistical Assessment of the Pollution Sources Along The Stream of Kamchia River, Bulgaria," Ecological Chemistry Engineering, Vol. 14, 2007, pp. 867-874.

15. D. L. Massart and L. Kaufman, "Interpretation of Analytical Chemical Data by the Use of Cluster Analysis," Journal of Wiley, New York, 1983.

16. B. Vandeginste, D. L. Massart, L. Buydens, S. De Long, P. Lewi and J. Smeyers-Verbeke, "Handbook of Chemometrics and Qualimetrics", Elsevier, Amsterdam, 1998.

17. J. W. Einax, K. H. Zwanziger and S. Geiss, "Chemometrics in Environmental Analysis," VCH Weinheim, 1998.

18. T. Kohonen, "Self-organizing Maps", Springer, Berlin, 2001.

Chapter 3

Assessing Domains in Quality of Life? A Case Study

Sayanti Sen[1], Goutam Sen[2]

[1]*Department of Geography, Bangabasi College, Calcutta University, Kolkata, India*

[2]*School of Oceonographic Studies, Jadavpur University, Kolkata, India*

ABSTRACT

Mining in India has developed haphazardly with a significant part of mining activities being informal or unauthorised. Open cast quarries, coal washeries, thermal power plants, coke-oven plants, etc., contribute to serious air and water pollution. The aim of this paper is to assess the QoL of residents living in Jharia coalfield. In this study, QoL domains are analysed considering the social, cultural, environmental, health and economic condition of the study areas. A path model has been developed to analyse the cascading effect on domains. The purpose of investigating path model with overall quality of life was to contribute to the subjective wellbeing literature by analyzing the predictors of OQoL in the sample population. Therefore, OQoL was used as the dependent variable of a regression analysis using four domains as variables and leveraging path analysis method to arrive at the causal model.

Keywords: Quality of Life, Path Analysis, Minning, Jharia

1. INTRODUCTION

Mining in India has developed haphazardly with a significant part of mining activities being informal or unauthorised mining. Open cast quarries, coal washeries, thermal power plants, coke-oven plants, etc., contribute to serious air and water pollution. [1] -[3] mentioned of issues confronting with achieving desired development for economic or social reasons and as a result, persisting degradation of health, safety and wellbeing has been affecting quality of life. The region continues to witness serious ecological imbalance and associated environmental problems [4] . Fire and subsidence due to atrocious mining practices are some of the reasons which make such mining areas vulnerable; these areas are of grave concern and these have directly attributed to unscientific and unplanned coal mining in these areas. The magnitude of the problem has increased over the years with growth of habitation over unsafe areas.

As documented a number of mining activities impact environmental, health and safety problems, each of which adversely impacts human quality-of-life [5] [6] have studied QoL to plan for overall community development. [7] - [10] studied quality of life indexing system of Bhowra mine area and West Bokaro complexes in Jharkhand to identify the degrading situation of inhabitants' lifestyle prevalent in the area. Human quality of life was studied for Bokaro mining complex by [11] and for Jharia coalfield by [12] and QoL was quantified on the basis of resident's economic capability. The impact of [13] mine fire and subsidence on QoL in Jharia coalfield area and provided rehabilitation and relocation plan for the region. The issues confronting today are faced with achieving desired development for economic or social reasons on one hand and safeguarding the environment and maintaining good QoL on the other [5] .

The aim of this paper is to assess the QoL of residents living in Jharia coalfield. In this study, QoL domains are analysed considering the social, cultural, environmental, health and economic condition of the study areas. An interpretation of QoL indicators and domains has been presented while measuring QOL. A path model has also been developed to analyse the cascading effect on other domains. The purpose of investigating path model with overall quality of life was to contribute to the subjective wellbeing literature by analyzing the predictors of OQoL in the sample population. Therefore, OQoL was used as the dependent variable of a regression analysis using four domains as variables and leveraging path analysis method to arrive at the causal model.

The study area lies in the heart of Damodar valley along the north of Damodar River in the eastern part of India as seen in Figure 1.

The coalfield is named Jharia, situated in the eastern part of DRB, extends for about 38 km in the east-west direction and a maximum of 18 km in the north-south direction covering an area of about 450 sq. km and is a representative area of the whole DRB [5] . Jharia coalfield, is located in Dhanbad district of Jharkhand state which lies between 23°37'3"N and 24°4'N latitude and between 86°6'30"E and 86°50'E longitude. Jharia situated in the eastern part of the coal field extends for about 38 km in the East West direction and 18 km in North South direction covering of about 450 km^2 [14] .

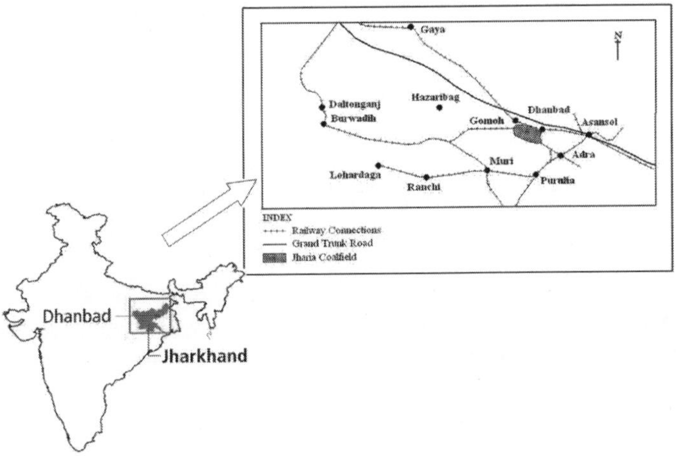

Figure 1. Study area.

2. METHODOLOGY

The methodological strategy chosen for this study is a case study approach and includes a combination of quantitative sources of evidence. The pre-tested self evaluating questionnaire was used to document the satisfactory levels of respondents residing in the core and buffer zones of the mining area. This questionnaire consisted of The sample survey was conducted covering about 70 households belonging to different caste-groups, occupation-groups and land size-groups to make it approximate to a stratified sampling method. Analysis of the acquired data has been done leveraging existing literature and validation using statistical techniques.

2.1. Interpreting Quality of Life

The survey was organized to collect information on key facets of home and environment variables at the household level. Besides making an analysis on aspects of living conditions and socioeconomic profile, environmental satisfaction, social and civil aspects of life satisfaction of the locality, an attempt has also been made to analyze the overall quality of life in the existing mining region.

For interpreting the quality of life of the study area, the data has been stratified by a 5-point scale of QoL assuming a normal distribution. The 5-point scale is defined as follows:

1) Significantly below Average QOL - values between $\mu - 3\sigma$ and $\mu - 2\sigma$.

2) Below Average QOL - values between $\mu-2\sigma$ and $\mu-\sigma$.

3) Average QOL - values between $\mu-\sigma$ and $\mu+\sigma$.

4) Above Average QOL - values between $\mu+\sigma$ and $\mu+2\sigma$.

5) Significantly above Average QOL - values between $\mu+2\sigma$ and $\mu+3\sigma$.

where μ = mean value and σ = standard deviation.

After stratifying the satisfaction score, a detailed analysis was done to investigate the cross-sectional factors to identify the sections of population whose QoL is directed by the parameters which are considered for assessing quality of life in various domains of life. The data have been categorically analyzed to detail the quality of life of the respondents in the study area. As part of the questionnaire a single question was asked to the respondents about an overall quality of life measure "Your Overall Quality of Life?".

2.2. Path Analysis for Overall Quality of Life (OQoL)

Having isolated four domains which influence the QoL, it was also found that a causal relationship needed to be established. As part of the questionnaire a single question was asked to the respondents about an overall quality of life measure "Your Overall Quality of Life?".

The Total QoL score for the respondents is an aggregation of the scores in the individual domains namely economic liberty, environmental liberty, community liberty and civil liberty. Many researchers in their work have impressed upon the need for measuring overall quality of life as a singular question to respondents and its correlation with the total quality of life score. As highlighted in the Table 1 the total quality of life score has strong correlation (r = 0.716) with the OQoL score.

Path analysis is a straight forward extension of multiple regression which aims to provide estimate of the magnitude and significance of hypothesized causal connections between sets of variables. The arrows indicate that the variable at the tail of the arrow affects the variable at the head of the arrow. The numbers on the arrows are the path coefficients and indicate the strength of the relationship. OQoL is the final dependent variable.

During the study multiple path models were developed and tested for correlation between the input domain parameters and the OQoL dependant variable. Figure 2 shows the evolution of path model. The equation below has been used to calculate the path coefficient:

$$Z_0 = P_{01}Z_1 + P_{02}Z_2 + P_{03}Z_3 + P_{04}Z_4$$

The path models were tested for correlation between the domains and the causal output of overall quality of life. Model 3 with causal relationship between economic liberty domain and other 3 domains showed the highest correlation value and the path coefficients are highlighted.

Table 1. Correlation among dependent variables.

Correlation	Domain Wise Quality of Life Score					OQoL
	Economic Liberty	Environmental Liberty	Community Liberty	Civil Liberty	Total QoL	
Economic Liberty	-	0.312	0.357	0.457	0.899	0.665
Environmental Liberty		-	0.226	0.214	0.590	0.557
Community Liberty			-	0.279	0.563	0.301
Civil Liberty				-	0.631	0.304
Total QoL					-	0.716

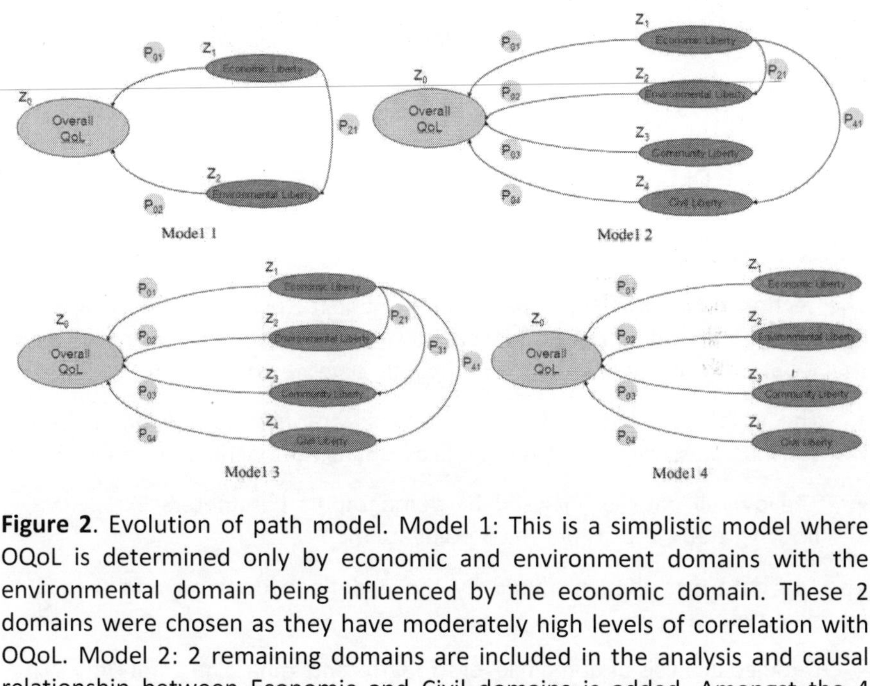

Figure 2. Evolution of path model. Model 1: This is a simplistic model where OQoL is determined only by economic and environment domains with the environmental domain being influenced by the economic domain. These 2 domains were chosen as they have moderately high levels of correlation with OQoL. Model 2: 2 remaining domains are included in the analysis and causal relationship between Economic and Civil domains is added. Amongst the 4 domains, Economic and Civil domains have the highest correlation. Model 3: Causal relationship between Economic and other 3 domains is hypothesized. Model 4: All domains are considered to be independent with either no correlation or with no causal relationship between them.

The overall effect on OQoL is a cumulative of direct effects and indirect effects. Each direct effect in the path diagram should be interpreted as the amount of change in a dependent variable due to a unit change in independent variable if no effects are transmitted through intervening variables. Indirect effects, on the other hand has been calculated and interpreted as the magnitude of the predicted change in the dependent variables through intervening variables if the independent variables is changes one unit. Indirect effect can be calculated by multiplying path coefficient for compound causal paths.

$$\text{Direct effect of economic liberty domain} = P_{01}$$

Indirect effect of economic liberty domain on OQoL is computed as a summation of the following components:

- $\text{economic} > \text{environment} > \text{OQoL} = P_{21}P_{02}$.
- $\text{economic} > \text{community} > \text{OQoL} = P_{31}P_{03}$.
- $\text{economic} > \text{civil} > \text{OQoL} = P_{41}P_{04}$.

$$\text{Hence indirect effect} = (P_{21}P_{02}) + (P_{31}P_{03}) + (P_{41}P_{04})$$

$$\text{Total effect} = \text{Direct effect} + \text{Indirect effect} = P_{01} + (P_{21}P_{02}) + (P_{31}P_{03}) + (P_{41}P_{04})$$

3. RESULTS AND DISCUSSIONS

The section represents relationship among the factors related to QoL to lay emphasis on providing better services to the locality and enhance the QoL of the people. The analysis was done in the following steps:

- The total respondent population was stratified based on the total satisfaction score by a 5-point scale assuming normal distribution.
- The overall data is analysed by demographic parameters like sex, age, housing status etc. as per the 5-point scale.
- The data was then analysed by the 4 defined domains and their sub-parameters.
- Detailed analysis of domain level strengths and impact on Total QoL.
- Correlation between the QoL domains, Total QoL and Overall QoL was determined.

3.1. Analysis of Quality of Life of Surveyed Population

For interpreting the QoL of the study area, firstly the mean score was calculated which is 83.97 and deviation from the mean is 13.48. The lower limit of QoL and higher limit of QoL was calculated subtracting the deviation from the mean to get the lower and higher limit. Figure 3 shows the spatial pattern of quality of life. The Table 2 summarizes the distribution of the respondents as per the 5-point scale.

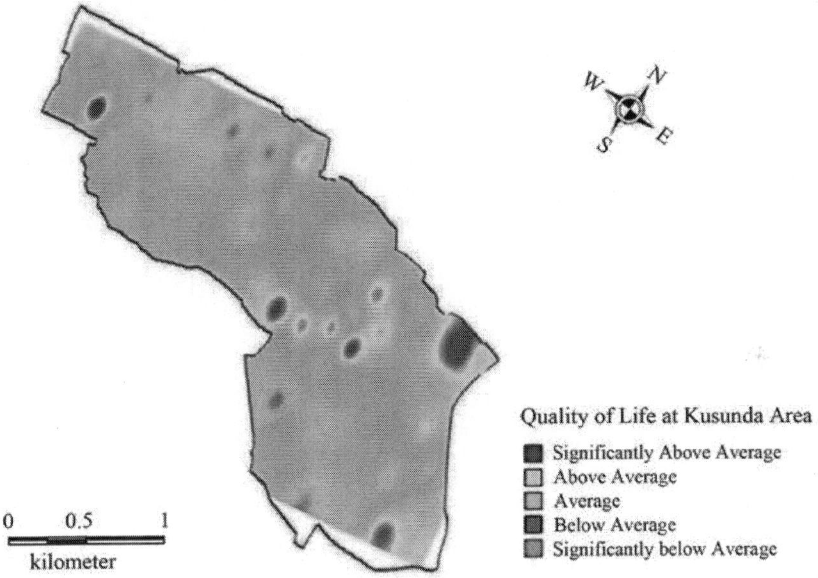

Figure 3. Quality of life at Kusunda area of Jharia Coalfields.

Table 2. Developing the 5-point scale for analysing quality of life scores using a normal distribution.

5-Point Scale	QoL Score		Distribution in (%)
	Lower Limit	Higher Limit	
Significantly below Average QoL	43.5	57.0	0.0
Below Average QoL	57.0	70.5	5.70
Average QoL	70.5	97.4	70.00
Above Average QoL	97.4	110.9	8.60
Significantly above Average QoL	110.9	124.4	5.70

3.2. Analysis of Quality of Life Based on Demographic Information

3.2.1. Sex Ratio

For this study the sample population was 70 among the sample respondents 10% females, and 90% interviews were males.

From Table 3, it is noted that none of the female respondents interviewed reported of having "Above Average" QoL. This might be attributed to the fact that the women population in the surveyed area are from economically backward parts of the society and have nominal influence upon the four liberty domains. 29% of female respondents reported of having "Below Average" QoL whereas 71% have an average QoL as shown in Figure 4.

Table 3. Quality of life by sex ratio.

Sex Ratio	Quality of Life				
	Significantly below Average	Below Average	Average	Above Average	Significantly above Average
Overall Population	0%	15.7%	70.0%	8.6%	5.7%
Female	0%	29%	71%	0%	0%
Male	0%	14%	70%	10%	6%

Detailed analysis showed that in economic liberty domain, women have satisfaction levels equally between "Below Average and 'Average' while in Environmental and community liberty domains the female respondents have an 'average' QoL. For civil liberty domain, they have a mixed satisfaction levels of QoL. It has been noted that life satisfaction perception response from male respondents have been ranging from Below Average to significantly Above Average. With maximum of 70% men having average QoL and 6% male respondents reported of having significantly Above Average QoL. Men with a 'Significantly above Average' in economic liberty domain have an overall QoL 'Significantly above Average'".

3.2.2. Age and Quality of Life

The demographic composition of sample population taken into consideration is distributed between the age group 25 years to more than 60 years. The age-group spread of the sample population is shown in the Figure 5 with 77% of the population covered between 25 - 50 age groups who are among the prime working groups.

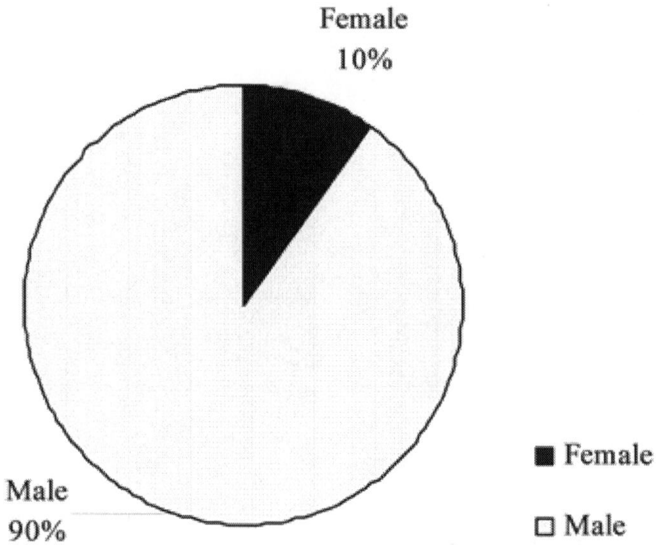

Figure 4. Age-sex ratio.

A further investigation about the QoL of the above respondents it reveals (Table 4) that 12% of the 31 - 40 age group respondents responded with a QoL "Significantly above Average" and in 41 - 50 age group, the "Significantly above Average" QoL population drops to 5%. The 25 - 30 and 41 - 50 age group responded with the highest "Average" scores of 78% and 79% respectively. The 51 - 60 age groups scored the maximum "Above Average" scores.

The analysis above corroborates the ground reality in India context where 25 - 30 age group struggles for a job and a living while 31 - 40 age group is where there is maximum assertion of youth in Economic, Community, Environmental and civil liberty domain areas and thereafter it degrades with increasing age due to socio-eco- nomic dependence on children and lack of social security norms.

Table 4. Age-wise distribution.

Age-Wise Distribution (in Years)	Quality of Life				
	Significantly below Average	Below Average	Average	Above Average	Significantly above Average
Overall Population	0%	15.7%	70.0%	8.6%	5.7%
25 - 30	0%	11%	78%	11%	0%
31 - 40	0%	23%	65%	0%	12%
41 - 50	0%	11%	79%	5%	5%
51 - 60	0%	8%	58%	33%	0%
>60	0%	25%	75%	0%	0%

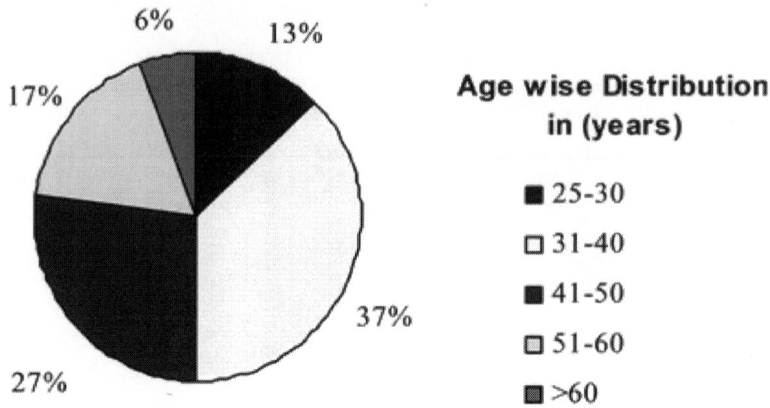

Figure 5. Age-wise distribution.

3.2.3. Land Ownership Status and Quality of Life

The respondents land ownership status and QoL were assessed keeping in mind the vulnerability of the study area due to mine fire and its economical viability making the region danger-prone for the residents to occupy. The study area is very much in news due to the fact that government at its different forums have been discussing for the evacuation of the area as a large portion is under fire, subsidence and inundation. The residents of the study area are not ready to evacuate the place as they earn their living from the surroundings which has developed into a good commercial hub. Three categories of ownership status ascertain in the study region. 54% of the respondents were found to be encroachers and the rest 44% were distributed among the legal land holders and tenants.

A further investigation about the QoL of the respondents land ownership it was seen that (Table 5) that 80% of the legal land holder have reported to have “Average” QoL while 10% each are “Below Average QoL” and “Above Average QoL”. 5% tenant and 8% encroacher are the only category of people flaunting to have “Significantly above Average” QoL.

Figure 6 shows the analysis above establishes the respondents who own a legal land holding are devoid of “Significantly above Average” QoL. It appears that with more security, as in case of ownership of land holding, comes an additional cost of maintaining it due to continues risk of subsidence, fire and fear of loosing the land probably negatively impacts QoL satisfaction levelTable 5 shows the variation below.

3.2.4. Housing Status and Quality of Life

In the mining region, a lot of land is illegally mined by private contractors. Besides, land is also converted into roads, townships for miners, infrastructure for administrative purposes, stockyard for preliminary processing operations, etc. Keeping with the inclination for illegal mining practices residents of the region, it was seen that 40% respondents have illegal housing, 46% rented housing and 14% owned houses. The booming coal based industrial economy has not only made the region a striking location for easy money making but also a lucrative region for several illegal activities. The housing pattern with 40% illegal housing is not a surprise.

The study area's socio-economics included an assessment of the coalfield area, a proposed rehabilitation strategy and a cost-benefit analysis for relocation and rehabilitation of affected people [15] .

With a history of subsidence, the region has been identified as one of the risk prone regions inundated with problems like land acquisitions, and an active underground coal mine fire. Such problems are caused due to haphazard growth of population resulting to such incompatible housing status. As shown in Figure 7 people own-

Table 5. Ownership status of land.

Ownership Status of Land	Quality of Life				
	Significantly below Average	Below Average	Average	Above Average	Significantly above Average
Overall Population	0%	15.7%	70.0%	8.6%	5.7%
Encroacher	0%	16%	66%	11%	8%
Legal Land Holder	0%	10%	80%	10%	0%
Tenant	0%	18%	73%	5%	5%

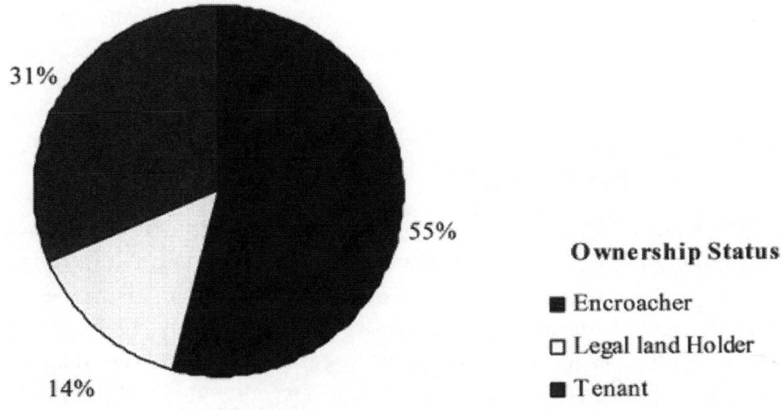

Figure 6. Land ownership status.

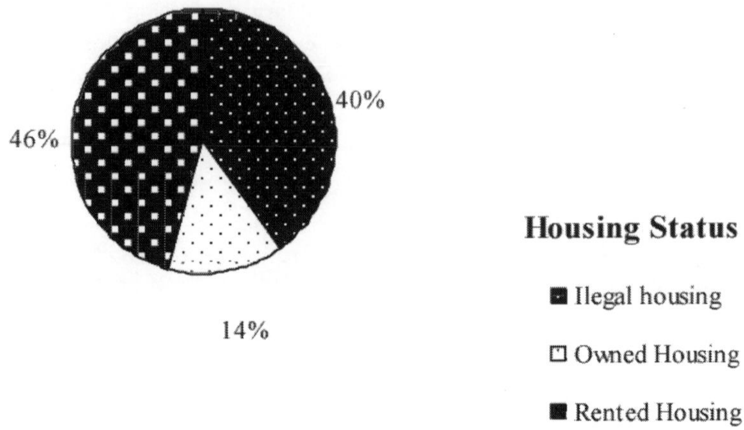

Figure 7. Housing status.

ing their housing do not have "Significantly above Average" QoL and may be attributed to higher cost of living. Table 6 shows However, 80% of the people owning houses reported "Average" QoL, the highest in all groups. The cost of living for rented yet legal housing is probably lower than for self owned housing and is a better option than illegal housing status.

3.2.5. Duration of Stay and Quality of Life

Human being does things which interests him and benefits the society. Hence, the impacts of the human activities, specially mining and associated activities, on the society assume a great importance on his stay in the community. The people of the mineral bearing areas, with the advent of mining and associated activities, are exposed to various developments and this tends to increase their aspiration for employment and better their well- being. It was noted from the sample of 70 respondents about 84% of the respondents are residents of the study area for more than 15 years.

Healthy human communities are an integral part of a society. The duration of stay of the population in the mineral rich industrialized community is a major driving factor to healthiness and well-being. From Table 7 and Figure 8 it is analysed that 100% of the respondents whose duration of stay is less than 10 years have Average QoL and their age group suggests that they have migrated to the study area potentially for an occupation. The percentage of population living between 15 - 20 years reports to have a high 78% population with "Average" QoL and 22% with "Below Average" QoL. We all know that the region have fashioned several big and small coal businessmen as well as illegal miners due to the regions capability to make coal into free flowing money.

Table 6. Housing status.

Housing Status	Quality of Life				
	Significantly below Average	Below Average	Average	Above Average	Significantly above Average
Overall Population	0%	15.7%	70.0%	8.6%	5.7%
Illegal Housing	0%	14%	64%	14%	7%
Owned Housing	0%	10%	80%	10%	0%
Rented Housing	0%	19%	72%	3%	6%

Table 7. Duration of stay in the community and quality of life.

Duration of Stay in the Community (in Years)	Quality of Life				
	Significantly below Average	Below Average	Average	Above Average	Significantly above Average
Overall Population	0%	15.7%	70.0%	8.6%	5.7%
<10	0%	0%	100%	0%	0%
10 - 15	0%	25%	38%	13%	25%
15 - 20	0%	22%	78%	0%	0%
>20	0%	14%	72%	10%	4%

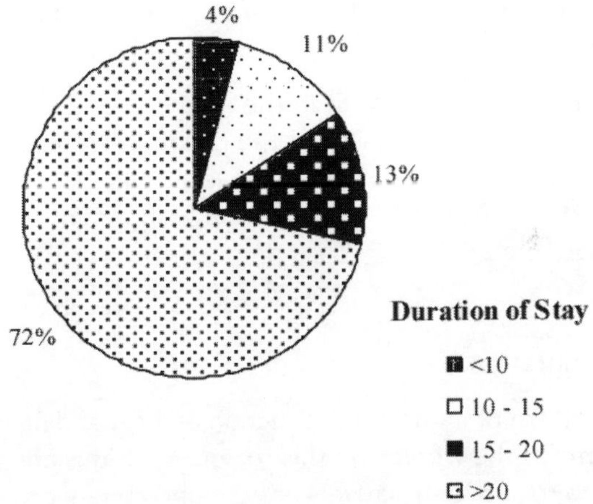

Figure 8. Duration of stay in the community.

Apparently the percentage of population living in this region for more than 20 years have the flavour of mining activities which have brought immediate capital to a region and restructured their activities and communities into a mining-dependent region. Consequently in this category, 4% the total population are reported to have significantly above average QoL, 10% above average QoL. Incidentally, 25% of the population staying between 10 - 15 years has significantly above average QoL.

3.2.6. Distribution of Occupation

The development of industrial and other associated activities in this area have increased the level of the economic activities creating several job opportunities; with labour supply has never been a problem for industrial expansion. Increased industrial and economic activities generate more money and increase the buying power of the people directly and indirectly associated with these activities. This leads to an increase in the cost of living, and it is believed that economic stability brings in satisfaction in human wants. The 1:2 ratios between the self employed (62%) and daily wage labour (33%) gives an idea regarding various job opportunities that reflect the ability to search work locally, types and nature of work available. In this region informal mining is an unofficial and unsanctioned often labelled as illegal mining. Percentage of self-employed can be an indicator of entrepreneurial drive of the workforce, but also be dominated by depressed and underemployed workers.

In regions like Jharia, the high self employed percentage also indicates that many people will act to improve their QoL, through elective occupational options. For example, if QoL, as measured by job availability, is improving in a region, it is likely more people are drawn to the area. Figure 9 shows the mobility of the work force and job growth in areas of high amenities are both factors in determining patterns of development in this region.

From the detailed analysis (Table 8) in the study area, occupational status has an important role in quantifying QoL of the residents. It is seen that daily wage labours and self employed respondents have a “Significantly above Average” QoL (4%) and (7%) respectively. 100% of the respondents who are employed have an “Average” QoL.

3.2.7. Level of Education

This changing pattern of industrial production and expanding work force has its own implications for the workers in this region. With this abundance of labour, in a resource rich region it is relatively easy to find men, women and children to take up ill-paid and unsafe jobs. The more deprived the social origin of the worker, the greater is the compulsion to take up any job irrespective of its hazards. Thus easy money from a hazardous job makes regions habitants reluctant towards better educational services. Here educational level was considered of importance because it is directly related to the quality of the workforce. Figure 10 shows 44% of the respondents are illiterate, only 31% of the respondents have primary level education, 17% have higher secondary level of education (Figure 10).

The QoL satisfaction level highly depends on the level of education and this is corroborated during this study as well. As seen in this is a potential explanation to the “Significantly above Average” QoL reported by a few respondents who are illiterate. The other set of "Significantly above Average" QoL are the people who have the highest level of education and this is noted to be the set of people also having "Significantly above Average" satisfaction level in Economic Domain. This confirms that education is indeed an enabler for economic prosperity and over all better QoL.

100% of the Undergraduate population have "Significantly above Average" QoL. 92% of the population who have studied up to higher secondary courses have "Average" QoL. 50% of the respondents having completed secondary school have "Above Average" QoL.

Predictably, the illiterate respondents show an n opposite phenomenon where 29% illiterate population have "Below Average" QoL and this group is what accounts for most of the overall "Below Average" QoL. Table 9 shows that respondents with no education tend to have lesser understanding and knowledge about the world beyond their location of stay and seem to be content with lesser attainments.

This is a potential explanation to the "Significantly above Average" QoL reported by a few respondents who are illiterate. The other set of "Significantly above Average" QoL are the people who have the highest level of education and this is noted to be the set of people also having "Significantly above Average" satisfaction level in Economic Domain. This confirms that education is indeed an enabler for economic prosperity and over all better QoL.

Table 8. Occupational status and quality of life.

Occupational Status	Quality of Life				
	Significantly below Average	Below Average	Average	Above Average	Significantly above Average
Overall Population	0%	15.7%	70.0%	8.6%	5.7%
Daily Wage Labour	0%	30.4%	60.9%	4.3%	4.3%
Employed	0%	0%	100.0%	0%	0%
Self Employed	0%	9.3%	72.1%	11.6%	7.0%

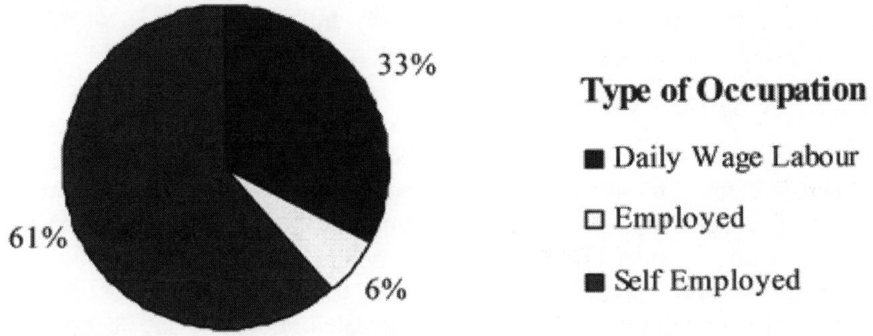

Figure 9. Distribution of occupation.

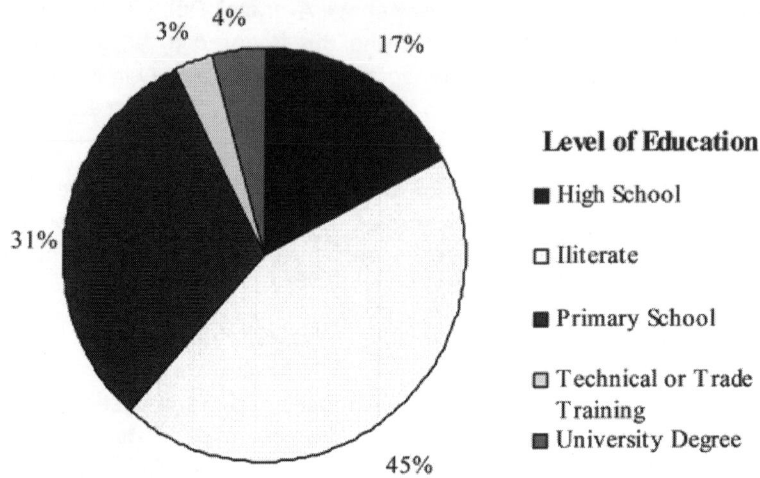

Figure 10. Distribution of occupation.

3.3. Analysis of Domain Based Quality of Life

QoL Domains are standardized measures with which to monitor and compare the QoL of a group or population. The approach taken in this paper was to select a list of formative indicators from each of the selected domains and was included in a QoL questionnaire.

One of the important outcomes of this research is the identification of QoL Predictors?those domains that seem to impact the perceptions of QoL in each community and for the respondents as a whole. Hence this section of the paper deals with individual domain and its correlation with Total QoL of the group of respondents.

A Heat Map is generated to visually depict the dependency of individual domain on the Total QoL. The X axis has the domain scores in the 5-point scale (SBA, BA, A, AA, SAA)[1] while Y axis contains the Average QoL score in the 5-point scale. Additionally, a simple regression equation has been developed demonstrating the correlation between the domain and the Total QoL.

3.3.1. Economic Liberty Domain

Economic issues such as income and wealth are significant in quantifying QoL. The prospect of a higher income has been identified as a major factor in motivating people to move to such communities [16] [17] . The economic domain includes aspects of one's monetary life and material wealth. This section focuses on the economic aspects of the respondents QoL. Domain questions asked about the adequacy of financial support and satisfaction with the amount of control he/she has over those financial resources in case of saving for the future. From the Figures 8-10, it shows less than 5% of the surveyed respondents have Average QoL in economic domain and reports to be having Below Average Total QoL. Higher the economic domain satisfaction ensures higher Total QoL. A small percentage of respondents rating themselves to be "Significantly above Average" in economic domain rate their Total QoL as "Above Average", "Average" and "Significantly Above Average". More than 50% of respondents having average satisfaction in economic domain have "Average" Total QoL and more than 5% respondents having "Below Average" economic liberty domain satisfaction have Average Total QoL score as shown in Figure 11.

Table 9. Level of education and quality of life.

Level of Education	Quality of Life				
	Significantly below Average	Below Average	Average	Above Average	Significantly above Average
Overall Population	0%	15.7%	70.0%	8.6%	5.7%
High School	0%	0%	92%	8%	0%
Illiterate	0%	29%	65%	3%	3%
Primary School	0%	9%	77%	14%	0%
Technical or Trade Tranning	0%	0%	50%	50%	0%
University Degree	0%	0%	0%	0%	100%

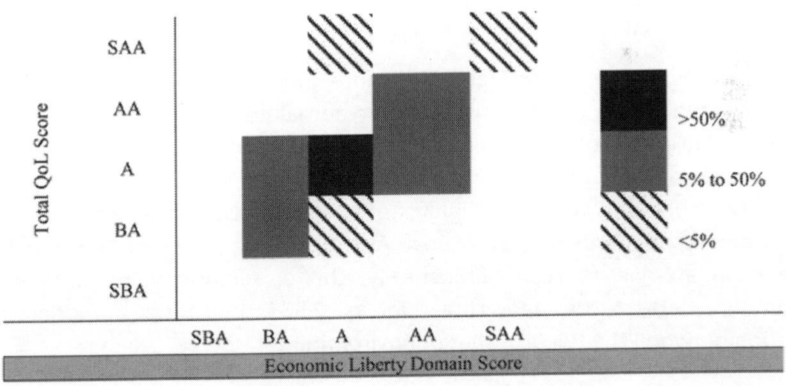

Figure 11. Heat map of economic liberty domain satisfaction with Total QoL.

Figure 12 the scatter diagram in the graph along with the regression equation above shows a strong correlation between respondents' satisfaction with life and financial security and stability. Statistically, the correlation value between Economic Liberty score and Total QoL is 0.899 indicating a strong correlation between the two. Higher the Economic score, higher would be the Total QoL.

3.3.2. Environmental Liberty Domain

The environment domain includes the following indicators: social security, development schemes, access to nature, cleanliness and security, basic infrastructure, services and goods available and transportation options. However, statistically, satisfaction with the environmental domain does not uniquely predict satisfaction with QoL. It is clear from the literature review, qualitative analysis and participant observation that environmental aspects are of critical importance to QoL in general [18] [19] .

Most of the participants in this research mentioned environmental qualities as a significant determinant of satisfaction with QoL in their region with more than 50% of the respondents who scores Average satisfaction in environmental domain also have "Average" Total QoL score. Figure 13shows the total quality of life in environmental domain.

As depicted in Figure 14 the heat map, there is a lot of scatter in environment scores and Total QoL scores e.g. there is a small population, less than 5%, for whom a "Significantly Average" Environment score does not result in more than "Average" Total QoL. Statistically, the correlation value between Environmental Liberty score and Total QoL is 0.590 indicating a moderate correlation between the two.

3.3.3. Community Liberty Domain

This section focuses on respondent's perceived ability to perform accustomed functions and social activities of daily living including relationship with neighbour, and related factors included for community living.

As shown in the Figure 15 and Figure 16, more than 50% of the respondents having "Average" score in community liberty domain also have Average Total QoL. The heat map scatter is similar to environment domain with lesser correlation e.g. a small population of less than 5% respondents having Average Community score but having "Significantly above Average" Total Score. Similarly few respondents who are Above Average in the scores of Community Liberty have an Average QoL. Statistically, the correlation value between Community Liberty score and Total QoL is 0.563 indicating a moderate correlation between the two like that of Environmental Liberty.

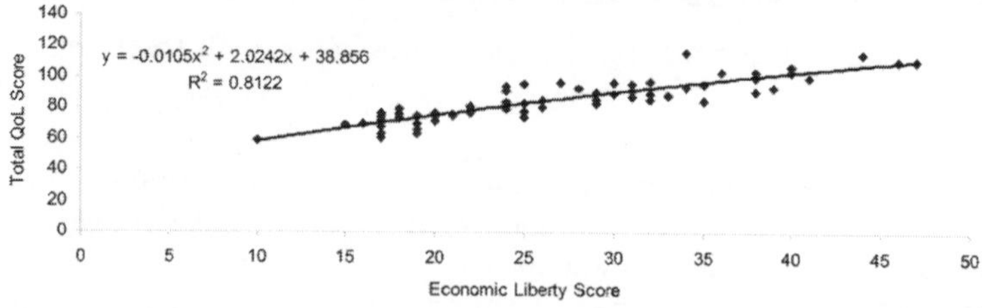

Figure 12. Scatter plot of total quality of life with economic liberty domain.

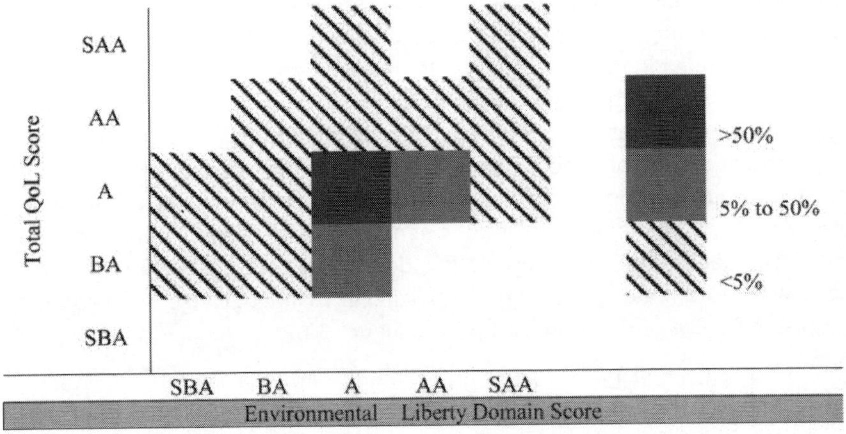

Figure 13. Heat map of economic liberty domain satisfaction with Total QoL.

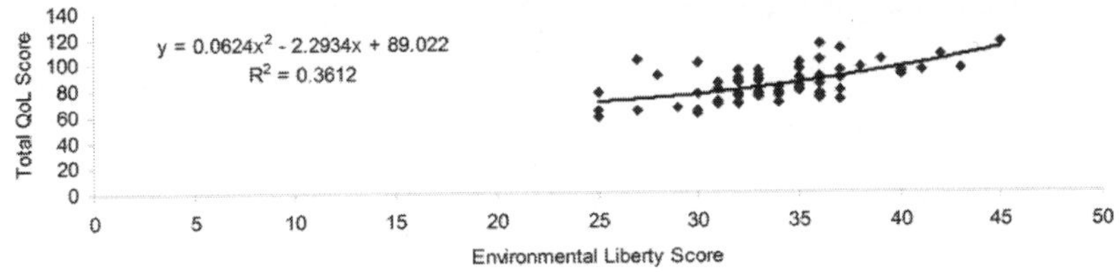

Figure 14. Scatter plot of total quality of life with environmental liberty domain.

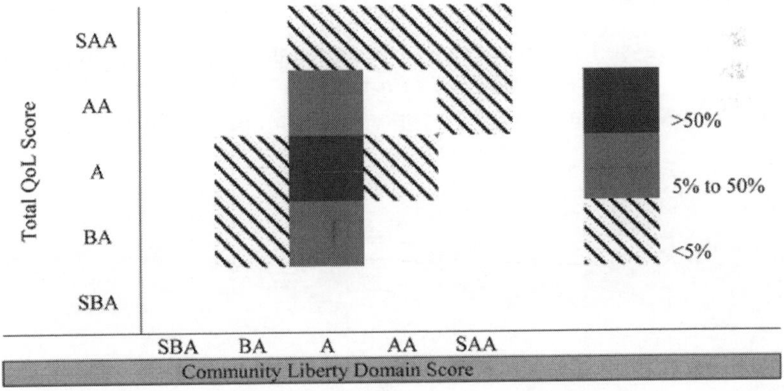

Figure 15. Heat map of community liberty domain satisfaction with Total QoL.

3.3.4. Civil Liberty Domain

The civil liberty domain includes indicator related to various facilities in the civil society. However, from the survey of the study area it is evident that the residents of the region who are deprived of services are not satisfied with the civil liberty domain. Human being are concerned about several categories which cover individuals' concern about income, assets, health, and leisure opportunities, while other categories are aimed at concerns typically entrusted to governments such as ensuring the vivacious condition of a community. The study population is from the deprived sections of the society struggling with basic needs and their needs correspond to the bottom three levels of Abraham Maslow's theory of Hierarchy of Needs namely Physiological, Safety and Social needs basic needs. Unlike the previous 3 domains, Civil Liberty scores are well spread with Total QoL scores with no particular Civil Liberty score accounting for more than 50% population as shown in Figure 17.

Figure 18 a scatter diagram in the graph along with the regression equation above shows a moderate correlation which is corroborated statistically with a correlation value of 0.631 between civil liberty score and Total QoL. This score, incidentally, higher than the correlation for Environmental and Community domains.

3.3.5. Correlation among the Four Quality of Life Domains

As shown in Table 10, although the correlation between the individual domains and Total QoL range from moderate to strong correlation, the inter-domain correlations are significantly weak. Economic liberty domain and civil liberty domain share the highest inter-domain correlation of 0.457.

4. DISTRIBUTION OF OVERALL QUALITY OF LIFE SCORE

Overall QoL is the self rated satisfactory score given by the respondents themselves when asked "How do they rate their life" on a Likeret scale of 0 - 10. Figure 19 shows the percentage of respondents rating their OQoL.

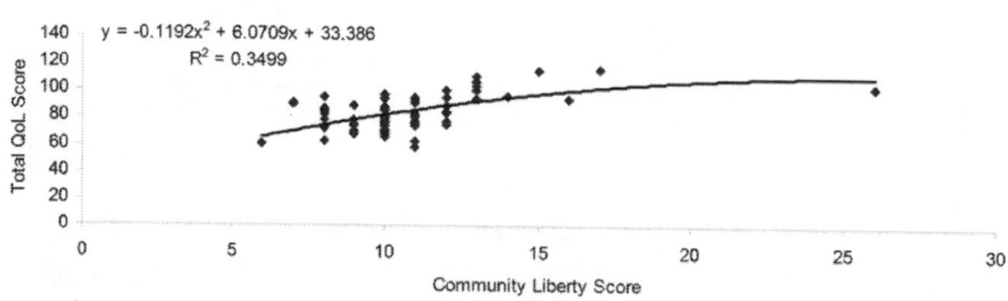

Figure 16. Scatter plot of total quality of life with community liberty domain.

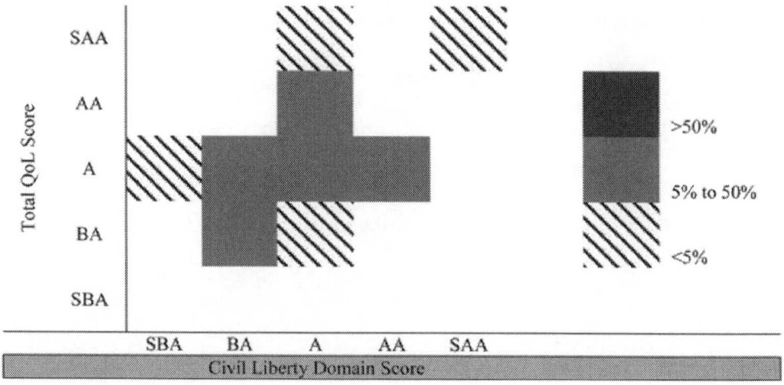

Figure 17. Heat map of civil liberty domain satisfaction with Total QoL.

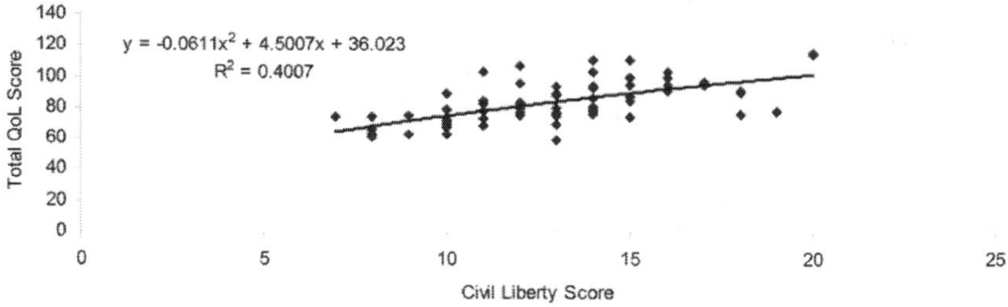

Figure 18. Scatter Plot of total quality of life with civil liberty domain.

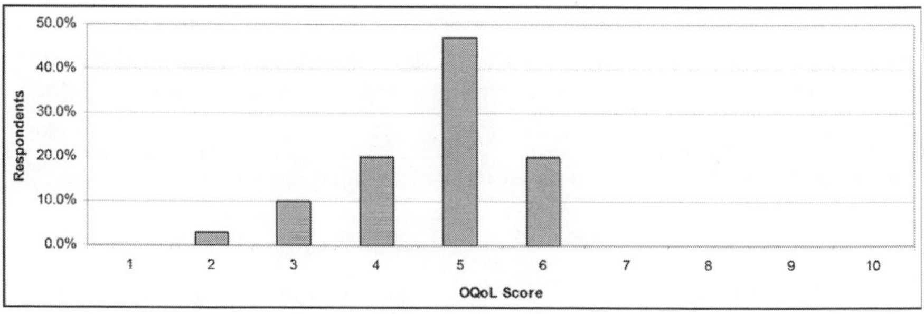

Figure 19. Overall quality of life score distribution.

Table 10. Correlation between domains and total quality of life scores.

Liberty Domain	Domain Wise Correlation Score				
	Economic	Environmental	Community	Civil	Total QoL
Economic	1.000	0.312	0.357	0.457	0.899
Environmental	-	1.000	0.226	0.214	0.590
Community	-	-	1.000	0.279	0.563
Civil	-	-	-	1.000	0.631
Total QoL Score	-	-	-	-	1.000

It may be noted that 67% of the population or two-thirds of the population had the view that they have an Overall QoL perception greater than or equal to a "Neutral" score of 5. The overall mean score for the population is 4.7. This is in contrast with the individual mean scores of the domains (economical liberty 2.184, environmental liberty 3.766, community liberty 2.163, and civil liberty 2.183). This demonstrates that although on a single question baseline of OQoL a person might have a better impression of his/her QoL but when questioned on deeper sub-elements which comprise the QoL, the individual domain scores dip lower than the overall impression.

The Table 11 shows the correlation between OQoL rated by the respondents with respect to individual QoL domains. The Total QoL score as a summation of the domain scores has a strong correlation with the OQoL as expected. Similarly, the Economic Liberty score shows the strongest correlation amongst the domains. Following it is the higher moderate correlation between the Environmental score and OQoL. For community and civil liberty domains there is a low correlation with OQoL. This demonstrates that people's perception about overall QoL is strongly driven by economic well-being and environmental well-being to an extent while community and civil matters are not strongly correlated.

Table 11. Correlation of domains score and overall quality of life.

Domains Score	Overall Quality of Life
Total QoL Score	0.716
Economic Liberty Score	0.665
Environmental Score	0.557
Community Liberty Score	0.301
Civil Liberty Score	0.304

Path Analysis for Overall Quality of Life Measure

Figure 20 displays a path analysis which shows a causal progression from top to bottom and then towards left. The numbers on the arrows are path coefficients and indicate the strength of the relationship. Overall satisfaction with life is the final dependent variable.

People's response indicates that the path coefficients derived from the selected model demonstrates at economic and environmental domains having a greater influence on the OQoL. The more important of the two is economic domain: a higher level of satisfaction on one's economic domain reflects upon the overall subjective QoL of the person. This observation is strengthened by the high correlation and moderately high path coefficient (0.518605).

The second is the environmental domain: the greater the quality of the environment which have been explained by having better social security in the community, cleanliness and quality of surrounding the respondents live in, availability of public utilities like market, bank, hotel facilities, communication facilities like post office, telephones, mobile connectivity and pollution level in the environment (air, water, noise, soil). Therefore greater the respondents are satisfied with the above conditions higher the overall life quality. This is substantiated by the moderate path coefficient (0.38707) among environmental liberty domain and overall QoL.

Community liberty has nominal impact on overall QoL with low path coefficient (0.07319). At the same time economic domain has greater impact on community liberty domain with 0.3570 as path coefficient. Therefore respondent's higher satisfaction towards economic domain reflects higher satisfaction towards over all life. Civil liberty is influenced by OQoL as demonstrated by the negative path coefficient between the civil liberty domain and the overall QoL. It also demonstrates how the economic liberty domain has an overarching influence on OQoL as also having indirect impact on OQoL through its influence on other domains. The OQoL score has a moderate correlation with economic and environmental liberty and a low moderated correlation with community and civil liberty. Within the domains there is a moderate correlation between economic and civil liberty while other correlations are significantly weak.

As indicated by the path coefficients findings from the model indicate that for the entire sample population, mostly the environmental and the economic domains were predictors of satisfaction with life as a whole. The relative impact of community liberty domain on the overall QoL is limited compared to economic and environment domains and this reflects the fact that education, health facilities, transport facilities mean little compared to economic liberty parameters like income, work, eating habits and environmental parameters. The path coefficient of civil liberty with OQoL is negative and signified a reverse causal impact i.e. higher OQoL usually would result in greater satisfaction with civil liberty parameters like political stability equal opportunities and relationship with neighbours.

The path model establishes the fact that economic upliftment of the area will enable greater satisfaction in the economic liberty domain. This has a cascading effect on other domains and this is reflected in the following results of direct and indirect effect of economic liberty domain on overall QoL.

$$\text{Direct effect of Economic Liberty domain on OQoL} = P_{01} = 0.5186$$

$$\text{Indirect effect of Economic Liberty domain on OQoL} = (P_{21}P_{02}) + (P_{31}P_{03}) + (P_{41}P_{04}) = 0.1467$$

$$\text{Total effect of Economic Liberty domain on OQoL} = \text{Direct effect} + \text{Indirect effect} = 0.6653$$

This shows that indirect effect of economic liberty through environmental, civil and community liberty domains is 22% which is significant. Further this means that every 1.0 standard deviation increase in economic liberty score leads to a 0.6653 standard deviation increase in OQoL. The correlation achieved between the path model regression equations and the OQoL has a strong correlation value of 0.7437 which is higher than the correlation derived from other path models used during the evolution to the final path model.

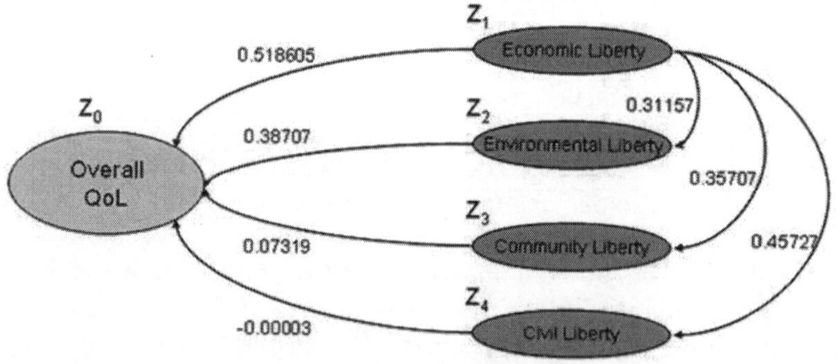

Figure 20. Path model of overall quality of life (OQoL).

These findings indicate that in order to impact satisfaction with life as a whole in the region, close attention must be given to economic and environment related aspects of life.

5. CONCLUSIONS

It was seen that 70% respondents in the JCR reported to have average QoL in the range of 2.20 to 3.05 for that region which is below the QoL Likert scale of 5 for “Satisfied”.

The opportunities in the study that exist today are indicative of the change in order to improve the QoL. Such opportunities include the changing economic and civil cultures in the region; strong environmental and community services; and academic institutions that are dedicated to the improvement of the biophysical, economic, and social conditions. However, these possibilities alone will not create a healthier, region. The domain wise approach towards accessing QoL and integrating environmental, social, economic liberty information is to better inform decision makers of the implications and benefits of alternate courses of action are compatible with each other. The main value of integration of indicators in these differing realms is to get a more complete picture of contemporary society.

The impact the mining on local people depends on the resource available to and conditions faced by the local people who follow mining activities. Mining activity has to ensure that there is improved QoL giving emphasis on the social, environmental and health impacts of mining and the implications of resource demand of mining vis-à-vis the rights and needs of the people.

The Path Model establishes the fact that economic upliftment of the area will enable greater satisfaction in the economic liberty domain. This has a cascading effect on other domains and this is reflected in the direct and indirect effect of economic liberty domain on OQoL. Path Analysis will helps us with establishing a hypothesis on the causal effect of the various parameters on OQoL. With the selected path model, it is established that Economic domain has the highest influence on OQoL while it also has a causal effect on Environmental and Community Domains. Environmental domain is the second most influencer of OQoL while civil liberty domain has a negative path coefficient indicating that OQoL influences Civil Liberty than the other way around. This shows that indirect effect of Economic Liberty through environmental, civil and community liberty domains is 22% which is significant. Further this means that every 1.0 standard deviation increase in Economic Liberty score leads to a 0.6653 standard deviation increase in OQoL. The correlation achieved between the path model regression equations and the OQoL has a strong correlation value of 0.7437 which is higher than the correlation derived from other path models used during the evolution to the final path model.

It can be concluded from the study that QoL is a multidimensional concept and the dimension that emerged in the form of domains like?conomic Liberty, Environmental Liberty, Civil Liberty and Community Liberty. The study helped to standardize the questionnaire tool to assess the QoL of residents of the problem area and also to suit the Indian scenario. Findings indicate that in order to impact satisfaction with life as a whole in the mining region, close attention must be given to economic and environment related aspects of life.

Illustration from the study area reveals details of the environmental and living conditions of the families. Government over years has tried to relocate affected families but have failed. In this research QoL analysis shows that the residents of this region though very satisfied with the economic facet of their individual life's but are highly dissatisfied with the environmental community facets of living. Looking in the vulnerable conditions several suitable rehabilitation and relief for coal-fire families have been planned. In one of the biggest rehabilitation plans in the world, the centre has announced to launch a scheme to relocate to safer areas some 79 thousands families whose houses face the threat of being sucked in by underground fires raging for the last 40 years (TOI, 2009).

CSR (Corporate Social Responsibility) for the local community has to be implemented. Coal India the major coal mining company and other industrial units should give more than 50% of their profit in CSR activities. Jharia coalfields industrial units are dominated by big industrial corporate houses both in private (Tata's) and public sectors like Bharat Coking Coal Company Ltd are integral part of CIL unit.

Sayanti Sen, Goutam Sen (10 km) radius have been taken into consideration for Community Development Programme. The implementation of CSR will have environmental benefits like greater material recyclability, better product durability and function ability, greater use of renewable resources and integration of environmental management tools into business plans, management standards and ecolabeling processes.

REFERENCES

1. Karunakaram, C. (1982) Role of Environmental Geology in Environmental Planning for Mining Operations. Proceedings of National Seminar on Minerals and Ecology, Delhi, 16-18 February 1982.

2. Michalski, S.R. (2004) The Jharia Mine Fire Control Technical Assistance Project: An Analysis. International Journal of Coal Geology, 59, 83-90.

3. Schori, A., Scrymgeour, A.H. and Munshi, P.L. (1997) Environment Management Plan for the Jharia Coalfield. Proceedings 14th Annual Meeting of the American Society for Surface Mining and Reclamation, (CD-ROM), Austin, 10-15 May 2004.

4. Areeparampil, M. (1996) Displacement Due to Mining in Jharkhand. Economic and Political Weekly, 31, 1524-1528

5. Sarkar, B.C., Mahanta, B.N., Saikia, K., Paul, P.R. and Singh, G. (2007) Geo-Environmental Quality Assessment in Jharia Coalfield, India, Using Multivariate Statistics and Geographic Information System. Environmental Geology, 51, 1177-1196.

6. Pal, A.K. and Kumar, U.C. (2005) Quality Of Life (Qol) Concept for the Evaluation of Societal Development of Rural Community in West Bengal, India. Asia-Pacific Journal of Rural Development, XV, 2.

7. Mishra, P.C., Mishra, B K. and. Tripathy, P.K. (2008) Socio-Economic Profile and Quality of Life Index of Sample Households of Mining Areas in Talcher and Ib Valley Coal Mines in Orissa. Journal of Human Ecology, 23, 13-20.

8. Rapant, S., Dietzová, Z. and Cicmanova, S. (2006) Environmental and Health Risk Assessment in Abandoned Mining Area, Zlata Idka, Slovakia. Environmental Geology, 51, 387-397.

9. Saxena, N.C., Pal, A.K., Prusty, B.K. and Kumar, P. (1998) Quality of Life Index of the Mining Area. Special Issue on Environment of the Indian Mining and Engineering Journal, Centre of Mining Environment, Indian School of Mines, Dhanbad, 15-18 July 1998.

10. Saxena , N.C. , Pal, A.K., Kumar, P. and Prusty, B.K. (1997) Quality of Living in Indexing for Mining Complexes in India. 2nd World Mining Congress, Katowice.

11. Kumar, P. (1996) Socioeconomic Profile of West Bokaro Mining Complex. M.Tech Dissertation, Indian School of Mines, Dhanbad.

12. Prusty, B.K. (1996) An Investigation into the Socio-Economic Profile of Bhowra Area of Jharia Coal-Fields. M.Tech Dissertation, Indian School of Mines, Dhanbad.

13. Chakraborty, T. and Ghosh, A.K. (2005) Need and Prospect of R&R in Jharia Coalfield. Conference on Technological Advancement and Environmental Challenges in Mining and Allied Industries in the 21st Century, Rourkela, 5-6 February 2005.

14. Government of India (1998) Carrying Capacity Based Developmental Planning of Damodar River Basin. Published in Alternative Scenario, Volume II, Sponsored by Ministry of Environment and Forest, New Delhi.

15. Michalski, S.R. (2004) The Jharia Mine Fire Control Technical Assistance Project: An Analysis. International Journal of Coal Geology, 59, 83-90.

16. Riffel, J.A. (1975) Quality of Life in Resource Towns. Urban Affairs, Ottawa.

17. Sibbel, A.M., Sibbel, J. and Goh, K. (2006) Fly-In, Fly-Out Operations—Strategies for Managing Employee Well- Being. Proceedings of the International Mine Management Conference, Australasian Institute of Mining and Metallurgy (Ausimm), Melbourne, 16-19 October 2006.

18. Takaro, T., Nakamura, K. and Watanabe, M. (2002) Urban Residential Environments and Senior Citizens' Longevity in Megacity Areas: The Importance of Walkable Green Spaces. Journal of Epidemiology and Community Health, 56, 913-918.

19. Sarvan, E. (2006) Life in The City—Innovative Solutions for Europe's Urban Environment. European Communities, Luxembourg .

Chapter 4

Higher-Order Statistics for Automatic Weld Defect Detection

Sara Saber, Gamal I. Selim

Department of Computer Engineering, Faculty of Engineering and Technology, Arab Academy for Science and Technology, Cairo, Egypt.

ABSTRACT

Image processing and image analysis are the main aspects for obtaining information from digital image owing to the fact that this techniques give the desired details in most of the applications generally and Non-Destructive testing specifically. This paper presents a proposed method for the automatic detection of weld defects in radiographic images. Firstly, the radiographic images were enhanced using adaptive histogram equalization and are filtered using mean and wiener filters. Secondly, the welding area is selected from the radiography image. Thirdly, the Cepstral features are extracted from the Higher-Order Spectra (Bispectrum and Trispectrum). Finally, neural networks are used for feature matching. The proposed method is tested using 100 radiographic images in the presence of noise and image blurring. Results show that in spite of time consumption, the proposed method yields best results for the automatic detection of weld defects in radiography images when the features were extracted from the Trispectrum of the image.

Keywords: High Order Statistics; Defect Detection; Radiographic Images; Non-Destructive Testing

1. INTRODUCTION

Non-Destructive Testing NDT has been considered as a simple laboratory curiosity but throughout the past Five decades, it was proved to be one of the essential tools in industry. Although radiography is one of the oldest techniques of NDT, it is still one of the most important Criteria in controlling weld joint in various industries. Furthermore, Welding can be defined as the process of combining metals using heat, pressure or both with or without a filler metal to produce localized union through fusion or recrystallization across the interface [1]. It is essentially used for several reasons such as the lack or the difficulty of a material to function solely, economic reasons, simplicity in usage as well as maintenance and repair. The quality of welding is relatively depends on the application to evaluate whether it'll be considered of a good quality or not, if it meets appearance requirements and continues indefinitely to do the job for which it is intended to do therefore, it is of a good quality.

Non Destructive Testing (NDT) methods are used for welding process inspection by examining the surface and subsurface of the weld and surrounding base material. Five basic methods are commonly used to examine welds; visual, liquid penetrate, magnetic particle, ultrasonic and radiographic [2]. Radiography can be carried out using X-rays, a radioactive gamma source, or a neutron source. In gamma radiography, a gamma beam is transmitted from the radiation source to penetrate the weld. Different levels of absorption take place in the defective part and the surrounding metal, depending upon the material used as shown in **Figure 1** [3]. A radiographic record of the transmitted energy, called the radiograph, is obtained. It is based on the different degrees of penetration of the radiant energy in the weld. When less absorption occurs owing to a thin section of metal, dark regions appear in the radiograph, where more penetration of the rays takes place. On the other hand, thicker areas or higher density materials absorb more radiation, and their corresponding images on the radiograph are lighter. The film is developed and treated with chemical components in a darkroom before the radiograph of the weld interior can be inspected clearly using an illumination box [2]. The film can be used for evaluating the weld area by a radiographic expert or can be digitized to the computer using a digital camera or scanner. The interpretation of the radiographic film can be done automatically or by a skilled operator. The automatic weld defect detection in radio graphy images has been studied by several researchers using various techniques and approaches [4-11]. Recently, Kasban et al. [12,13] used Cepstral approaches depending on extracting the features from the transform domains or from the power density related to the images.

The rest of this paper is organized as follows. Section 2 presents the proposed automatic weld defect detection method. Section 3 presents the image enhancement procedure. The method of selecting the welded area is presented in Section 4. Section 5 presents the usage of Higher Order Statistics (HOS) in the automatic detection of weld defects. Finally, conclusions are illustrated in Section 6.

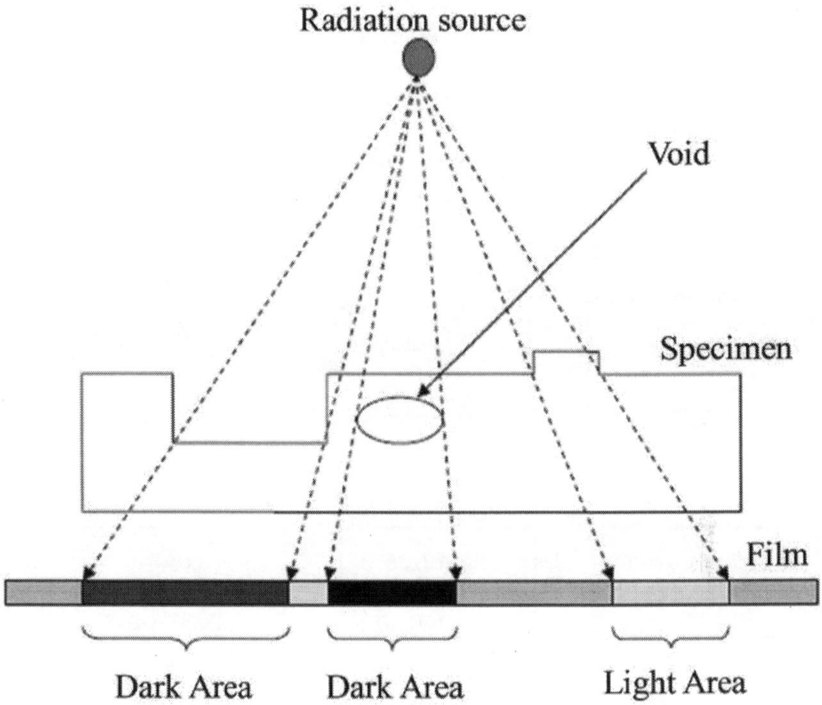

Figure 1. Gamma radiography principles.

2. PROPOSED AUTOMATIC WELD DEFECT DETECTION METHOD

The proposed approach in this paper is shown in **Figure 2**, the radiographic images were enhanced using Adaptive Histogram Equalization and filtered using Mean and Wiener filters. Secondly, the welding area was selected from the whole radiography image. Thirdly, the Cepstral features were extracted from the High Order Spectra (HOS) using Bispectrum and Trispectrum methods. Finally, neural networks were used for feature matching.

3. IMAGE ENHANCEMENT

Image enhancement is required for improving the contrast between the image background and the weld area and for removing any noise originating from the digitizing process. In this paper, the radiographic images were enhanced using contrast enhancement and image filtering.

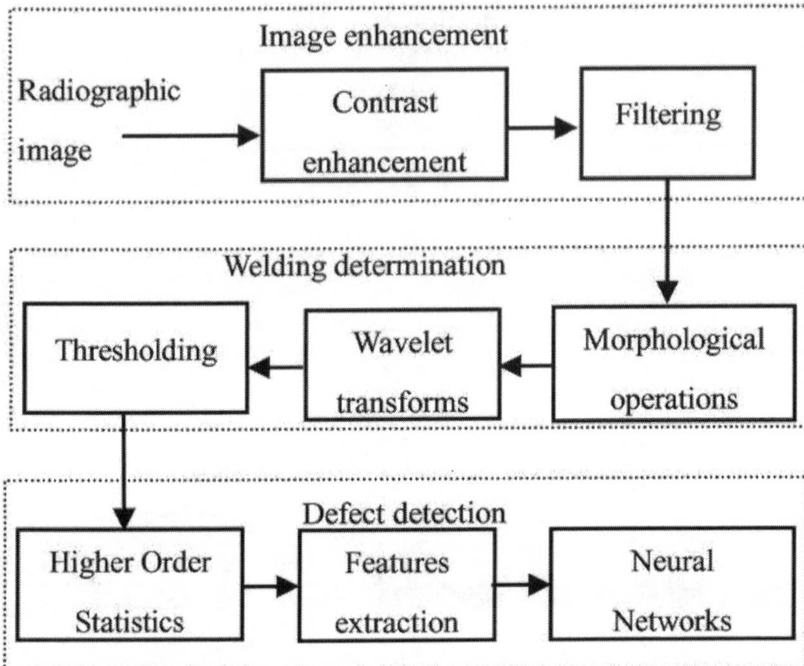

Figure 2. Automatic weld defect detection method.

3.1. Contrast Enhancement

The histogram equalization improves the visual appearance of the image by assigning equal number of pixels to all available intensity values [14]. Image contrast can be maximized using adaptive histogram equalization by adaptively enhancing the contrast of each pixel relative to its local neighborhood. This process produces improved contrast for all levels of contrast in the original image. In this process firstly, the histograms are calculated for small regional areas of pixels, producing local histograms, then equalize these local histograms or are remapped from the often narrow range of intensity values indicative of a central pixel and its closest neighbors to the full range of intensity values available in the display. Adaptive Histogram Equalization (AHE) can provide better contrast in local areas than that achievable utilizing traditional histogram equalization method. The traditional methods process the entire image at once, where adaptive histogram equalization utilizes local contextual region. Consider an image of size M × N, gray levels Gy, the cumulative histogram H(g), then the transfer function $F_{HE}(g)$ that maps the original gray levels into the transformed ones is define as [13]:

$$F_{HE}(g) = Gy_{\min} - (Gy_{\max} - Gy_{\min})\frac{H(g)}{MN} \tag{1}$$

where Gy_{max} and Gy_{min} are the upper and lower limits of the transformed gray values, respectively. The adaptive histogram equalization is made adaptive by taking the histogram over a local region instead of the whole image as [13]:

$$F_{\mathrm{AHE}}(g) = Gy_{\min} + (Gy_{\max} - Gy_{\min})\frac{H_{\mathrm{AHE}}(g)}{(MN)_{\mathrm{AHE}}} \tag{2}$$

3.2. Filtering

Image filtering can be used for removing the noise, sharpen contrast, or highlight contours from the radiographic images. Recently many researchers used and compared many filters and they proved that Wiener filter is the most suitable one for the radiographic image filtering process [13,15,16], so that wiener filter has been selected in this paper. Wiener filter is designed by minimizing the MSE between the filtered image and the original image. It can be applied to the image adaptively, tailoring itself to the local Image variance. Where the variance is large, it performs little smoothing, where the variance is small it performs more smoothing. Since the intensities of the pixels in the weld area distribute more as a Gaussian distribution than other areas in the image of weld [5], it was selected for filtering radiographic images of welds. **Figure 3** shows the effects of the image enhancement process on the radiographic image.

The results show that the contrast enhancement and the image filtering improve the radiographic image quality by improving the contrast between the image background and the weld defect regions.

4. WELDING AREA DETERMINATION

Determination the welding area from the hall radiography image is important process for elimination of irrelevant information, leaving only the area of interest in the image. Many researchers have been carried out this process using different algorithms and different data set and they achieved different results [13,16-17]. In this paper, the welding area is determined in three steps; the to the image using a flat disk structuring element. The second step is applying discrete wavelet transform (DWT) to the closed image which leads to a decomposition of this image into four components; the approximation component and three detail components. The last step is thresholding the image using a threshold value. This value is automatic obtained from the occurrence histogram. The occurrence histogram of the local variance image exhibits at least two distinct peaks. The largest peak corresponds to the background while the others correspond to objects in the image. The gray level corresponding to the first trough between the first two peaks is considered the optimal threshold value for discrimination between the objects and the background [13]. **Figure 4** shows the result of this process when applied on the enhanced image shown in**Figure 3**(c).

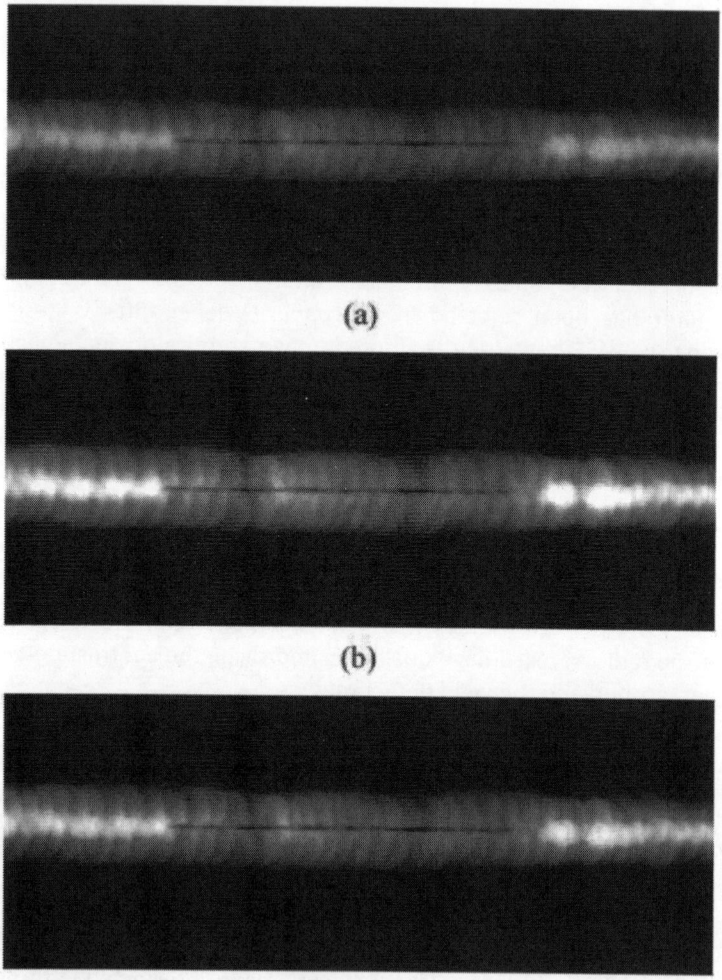

Figure 3. Radiographic image enhancement. (a) Original image; (b) After contrast enhancement; (c) After filtering.

5. WELD DEFECT DETECTION

In the proposed method for weld effect detection the Cepstral features are extracted from the HOS (Bispectrum and Trispectrum) related to the welding area from the radiographic images. These areas are lexicographically ordered into 1-D signals. The features are extracted from database consists of 100 signals. These features are used to train a neural network resulting features database. In neural network testing, the features are extracted from these signals then matching these features with the features that stored in the database during the training. For testing the proposed approach the tested signals have been degraded by different types of noise.

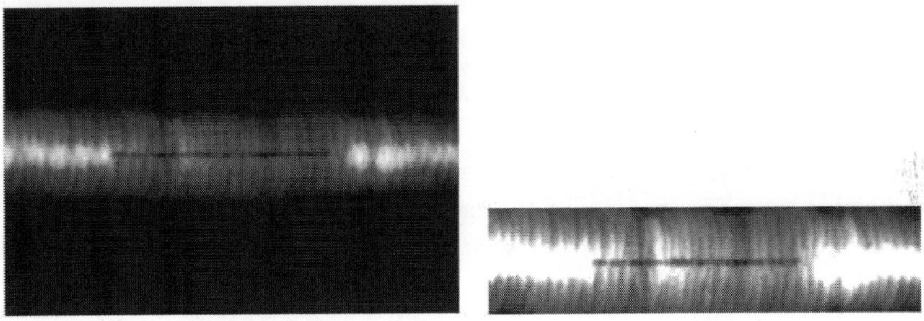

(a) (b)

Figure 4. Welding area determination. (a) Whole enhanced image; (b) Welding area.

5.1. Power Density Spectrum and Higher Order Statistics

Power Density Spectrum (PDS) describes how the power of a signal is distributed with frequency. There are several methods for estimating the PDS, which are divided into three categories; nonparametric methods, parametric methods and advanced-analysis methods [18]. In this paper we will take one method from the third category that has been achieved the best results in weld defect detection (MUSIC method) previously in [13]. The HOS are extensions of second-order measures such as the autocorrelation function and power spectrum. The Bispectrum provides an integral representation of the thirdmoment function of the process (skewness), where Trispectrum provides an integral representation of the fourth moment (kurtosis) [19]. The HOS are used because the second-order measures contain no phase information, so that the non-minimum phase signals cannot be correctly identified by 2nd-order techniques. And any Gaussian signal is completely characterized by its mean and variance. Consequently the HOS of Gaussian signals are either zero (e.g. the 3rd order moment of a Gaussian signal is zero), or contain redundant information. Many signals encountered in practice have non-zero HOS, and many measurement noises are Gaussian, and so in principle the HOS are less affected by Gaussian background noise than the 2nd order measures. E.g. the power spectrum of a deterministic signal plus Gaussian noise is very different from the power spectrum of the signal alone. However the Bispectrum of the signal and the noise is, at least in principle, the same as that of the signal [20]. HOS can be estimated using non-parametric or parametric methods. The non-parametric methods may be direct or indirect methods. The indirect method based on estimating the cumulant function and then taking Fourier transforms, while the direct method based on a segment averaging approach. The estimation methods in details are found in [21].

5.2. Feature Extraction

The image features are the most basic characteristics, it used to distinguish an image from others. The Mel Frequency Cepstral Coefficients (MFCCs) and the polynomial coefficients are used successfully in weld defect detection from radiographic images [12,13]. The Melscale is a mapping between the real frequency scale in Hz (f_{linear}) and the perceived frequency scale in Mels (f_{mel}). The Mapping is virtually linear below 1 kHz and logarithmic above as given by the following equation [12]:

$$f_{\text{mel}} = 2595\log_{10}\left(1+\frac{f_{\text{linear}}}{700}\right) \tag{3}$$

The mel-frequency warping is performed using a melfilter bank composed of a set of band-pass filters with constant bandwidths and spacing on the mel-scale .The bank consists of one filter for each desired mel-frequency component, where each filter has a triangular filter bandpass frequency response. The triangular filters are spread over the entire frequency range from zero to the Nyquist frequency. The number of filters is one of the parameters, which affect the recognition accuracy of the system. The last stage involves performing a DCT on the log of the Mel-spectrum. If the energy of the mth mel-filter output is $\tilde{C}(m)$ the MFCCs will be given as follow [12]:

$$C_j = \sqrt{\frac{2}{N_{mf}}}\sum_{m=0}^{N_{mf}}\log\left(\tilde{C}(m)\right)\cos\left(\frac{j\pi}{N_{mf}}(m-0.5)\right) \tag{4}$$

where $j = 0,1,\cdots, J_{mf}-1$, J_{mf} is the number of MFCCs, N_{mf} is the number of Mel-filters and c_j are the MFCCs. The number of the resulting MFCCs is 13, since most of the signal information is represented by the first few coefficients. The 0th coefficient represents the average log energy of the frame.

The MFCCs are sensitive to mismatches and time shifts between training and testing data. Thus, there is a need for other coefficients to be added to the MFCCs to reduce this sensitivity. Polynomial coefficients can be used for this purpose as these coefficients can help in increasing the similarity between the training and the testing signals. To calculate the polynomial coefficients, the time waveforms of the Cepstral coefficients are expanded by orthogonal polynomials. The following two orthogonal polynomials have been used [12]:

$$P_1(i) = i - 5 \tag{5}$$

$$P_2(i) = i^2 - 10i + 55/3 \tag{6}$$

To model the shape of the MFCCs time functions, a nine elements window at each MFCC is used. Based on this window assumption, the polynomial coefficients can be calculated as follow [12]:

$$a_j(t) = \frac{\sum_{i=1}^{9} P_1(i) c_j(t+i-1)}{\sum_{i=1}^{9} P_1^2(i)}, \quad b_j(t) = \frac{\sum_{i=1}^{9} P_2(i) c_j(t+i-1)}{\sum_{j=1}^{9} P_2^2(i)} \tag{7}$$

where $a_j(t)$ and $b_j(t)$ are the slope, and the curvature of c_j in the tth frame. The vectors containing all c_j, a_j and b_j are concatenated to form a single features vector.

5.3. Artificial Neural Network

The Artificial Neural Networks (ANNs) are computational representations based on the biological neural architecture of the brain [22]. ANN has several advantages such as; it can perform tasks more than the linear system, when an element of the ANN fails, it can continue without any problem by their parallel nature, ANN learns and does not need to be reprogrammed, ANN can be implemented in any applications, and finally, ANN can be implemented without any problem. The ANN suffers from some drawbacks such as; it needs training to operate, the architecture of ANN is different from the architecture of microprocessors therefore needs to be emulated, and it requires high processing time for large neural networks. The ANNs have been successfully applied in weld defect detection from radiographic images [12,23,24]. In this paper the ANNs are trained about the features of the radiographic images in the training stage. In testing stage it make a feature matching between the features of the tested image and the features that stored during the training. The used ANN is a Multi-layer perceptions (MLPs) neural network has an input layer, a single hidden layer, and an output layer. A single neuron only of the output layer is shown for simplicity. Each neuron in the neural network is characterized by an activation function and its bias, and each connection between two neurons by a weight factor. In this paper, the neurons from the input and output layers have linear activation functions and hidden neurons have the following sigmoid activation function:

$$F(u) = 1/\left(1 + e^{-u}\right) \tag{8}$$

Therefore, for an input vector E, the neural network output vector Y can be obtained according to the following matrix equation [13]:

$$Y = W_2 * F\left(W_1 * E + B_1\right) + B_2 \tag{9}$$

where W_1 and W_2 are the weight matrices between the input and the hidden layer and between the hidden and the output layer, respectively, and B_1 and B_2 are bias matrices for the hidden and the output layer, respectively. The training algorithm adapts the weights by attempting to minimize the sum of the squared error between a desired output and the actual output of the output neurons given by [13]:

$$E_o = \frac{1}{2} \sum_{o=1}^{O} \left(D_o - Y_o\right)^2 \tag{10}$$

where D_o and Y_o are the desired and actual outputs of the oth output neuron. O is the number of output neurons. Each weight in the neural network is adjusted by adding an increment to reduce E_o as rapidly as possible. The adjustment is carried out over several training iterations until a satisfactorily small value of E_o is obtained.

5.4. Weld Defect Detection Results

In radiographic systems, there is a variety of imaging noise sources, which originate from the different stages and elements of the system, such as gamma ray source, scattered radiation, imaging screen or electronic circuits among others [25]. The proposed method is run 8 times according to image degradations and image blurring. The degradations considered are Gaussian, impulsive, speckle and Poisson noise with and without image blurring. For each time, the features are extracted separately and compared from 5 signals related to the image. The first signal is the power density spectrum (PDS) of the image estimated by Multiple Signal Classification (MUSIC) method. The second signal is the Bispectrum of the image estimated by direct method. The third signal is the Bispectrum of the image estimated by indirect method. The fourth signal is the Bispectrum of the image estimated by ARMA method. The last signal is the Trispectrum of the image estimated by ARMA method.

Gaussian noise is the most common type of noise that results from the contributions of many independent signals [26]. In radiographic images, Gaussian noise may come from many natural sources, such as the thermal vibrations of atoms in conductors, black body radiation from the earth and other warm objects. To test the proposed method in the presence of the Gaussian noise, the percentage of detection rate is plotted with respect to the Signal to Noise Ratio (SNR) in decibel as shown in **Figure 5**. The results show that the detection rate is increase by increasing the SNR, and the parametric methods for higher order spectra estimation give higher detection rate than the nonparametric (direct and indirect) methods or PDS estimated by MUSIC method. The Bispectrum gives higher detection rate at low SNR (lower than 12 dB), and at high SNR the Trispectrum gives the higher detection rate.

Image blurring makes the image unfocused. It has been carried out by convolving the image with a low pass filter with kernel:

$$H_b = \frac{1}{25}\begin{bmatrix} 1 & 1 & 1 & 1 & 1 \\ 1 & 1 & 1 & 1 & 1 \\ 1 & 1 & 1 & 1 & 1 \\ 1 & 1 & 1 & 1 & 1 \\ 1 & 1 & 1 & 1 & 1 \end{bmatrix} \tag{11}$$

In the presence of the Gaussian noise with blurring, the detection rate is plotted against the SNR as shown in **Figure 6**.

The results show that the parametric methods for higher order spectra estimation give higher detection rate than the nonparametric or PDS estimated using MUSIC method. The Bispectrum gives higher detection rate at low SNR (lower than 10 dB), and at high SNR the Trispectrum gives the higher detection rate. By comparing the results in Figures 5 and 6 we see that the image blurring decrease the detection rate.

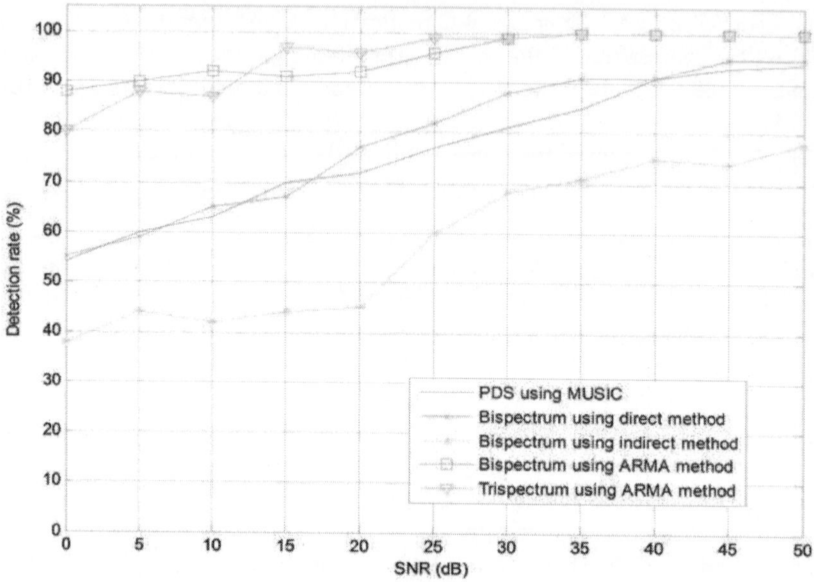

Figure 5. Detection results when the tested images are degraded by Gaussian noise.

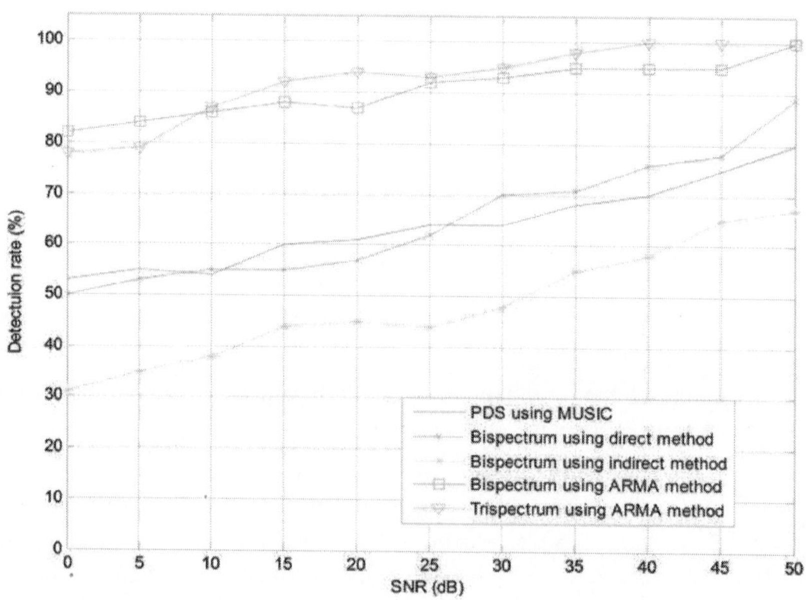

Figure 6. Detection results when the tested images are blurred and degraded by Gaussian noise.

Impulse noise corruption is very common in digital images. It is always independent and uncorrelated to the image pixels and is randomly distributed over the image [27]. Failures in sensors, readout circuits, A/D converters, or communication channels may introduce impulsive noise in digital images, in particular in radiography. Impulsive noise has been added to the radiographic images in the form of salt and pepper noise. In this case, the percentage detection rate is plotted against the percentage impulsive noise, as shown in **Figure 7**.

The results show that, the detection rate is decrease by increasing the impulsive noise, and the parametric methods for higher order spectra estimation give higher detection rate than the nonparametric (direct and indirect) methods or PDS estimated by MUSIC method. The Trispectrum gives the higher detection rate. In the presence of the impulsive noise after image blurring the percentage detection rate is plotted against the percentage impulsive as shown in **Figure 8**.

The results is similar to the results obtained in **Figure 7**, but it is noted that the image blurring decrease the detection rate. Speckle noise may be results from random fluctuations in the return signal from an object that is no bigger than a single image-processing element. It increases the mean grey level of a local area. Speckle noise has general model with the equation [28]:

$$G_{sn}(i,j) = F_{sn}(i,j)U_{sn}(i,j) + \xi_{sn}(i,j) \qquad (12)$$

where G_{sn}, F_{sn}, U_{sn} and ξ_{sn} refers to the observed image, original image, multiplicative component and additive component of the speckle noise respectively. In the presence of the speckle noise, the detection rate is plotted against the noise variance as shown in **Figure 9**.

The results show that, the detection rate is decrease by increasing the noise variance, and the parametric methods for higher order spectra estimation give higher detection rate than the nonparametric (direct and indirect) methods or PDS estimated by MUSIC method. The Trispectrum gives the higher detection rate. In the presence of the speckle noise after image blurring the percentage detection rate is plotted against the percentage impulsive as shown in **Figure 10**.

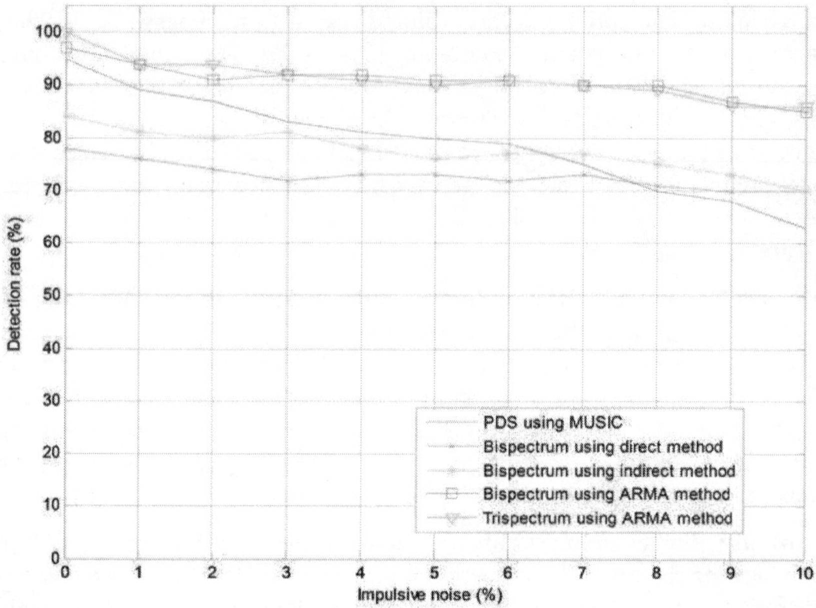

Figure 7. Detection results when the tested images are degraded by impulsive noise.

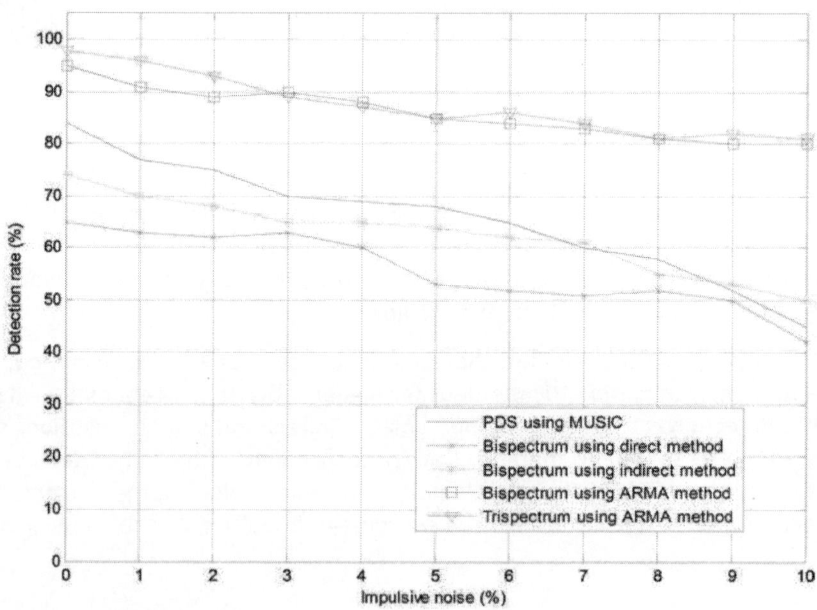

Figure 8. Detection results when the tested images are blurred and degraded by impulsive noise.

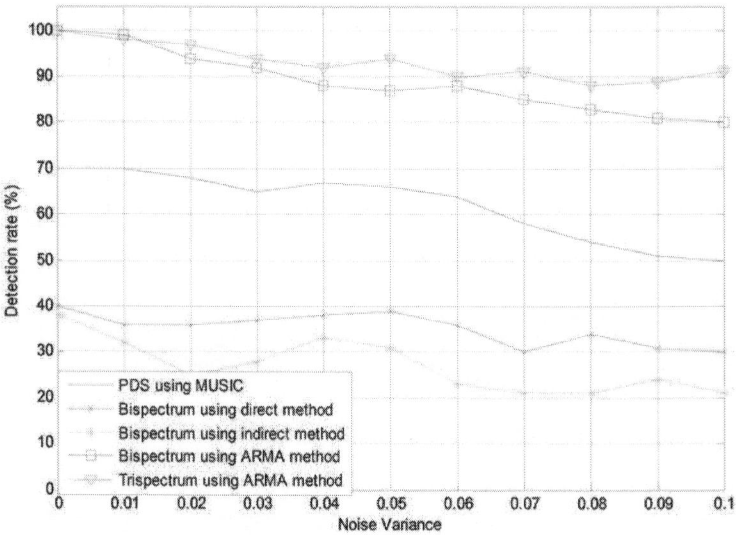

Figure 9. Detection results when the tested images and degraded by speckle noise.

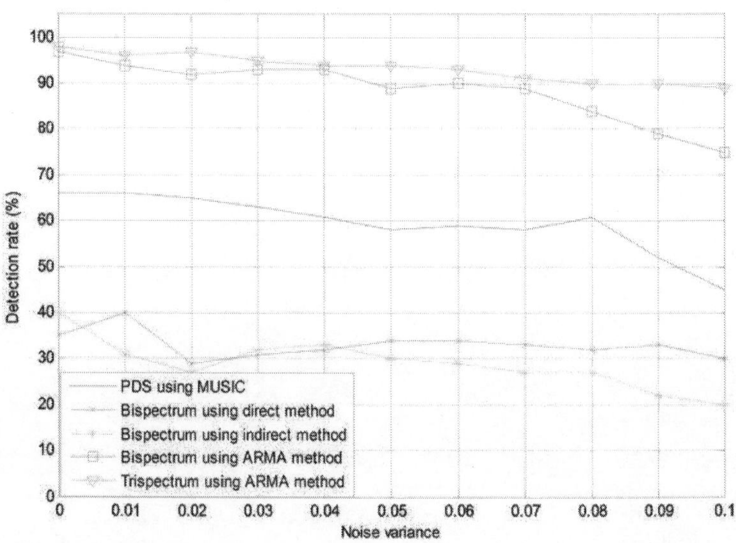

Figure 10. Detection results when the tested images are blurred and degraded by speckle noise.

The results is similar to the results obtained in **Figure 9**, but it is noted that the image blurring decrease the detection rate. Unlike Gaussian noise, Poisson noise is proportional to the underlying signal intensity, which makes separating

Table 1. Detection results when the tested images are degraded by Poisson noise with and without blurring.

Transform	Detection rate %	
	Without image blurring	With image blurring
PDS estimated MUSIC method	81	75
Bispectrum estimated by direct method	73	72
Bispectrum estimated by indirect method	66	61
Bispectrum estimated by ARMA method	97	95
Trispectrum estimated by ARMA method	98	98

Table 2. The processing time for testing 100 images by the proposed method for different contaminations.

	Degradation of tested image	PDS MUSIC	Bispectrum			Trispectrum
			direct method	indirect method	ARMA	ARMA
Train time (Sec)		135	163	179	434	470
Test time (Sec)	Gaussian noise	50	6890	7304	7756	8357
	Gaussian noise with blurring	53	7487	7789	7945	8111
	Impulsive noise	52	6251	7078	7970	8457
	Impulsive noise with blurring	55	6905	7139	8274	8945
	Speckle noise	56	6303	6464	7409	7930
	Speckle noise with blurring	57	6455	6580	6900	7571
	Poisson noise	50	6308	6542	6739	7663
	Poisson noise with blurring	51	6620	6780	6972	7061

signal from noise a very difficult task [29]. It may be result when the finite number of particles that carry energy is sufficiently small so that uncertainties due to the Poisson distribution, which describes the occurrence of independent random events, are of significance. In the presence of the Poisson noise with and without blurring, the detection rates are summarized in **Table 1**.

The results show that, the parametric methods for higher order spectra estimation give higher detection rate than the nonparametric (direct and indirect) methods or PDS estimated by MUSIC method. The Trispectrum gives the higher detection rate. The major disadvantage of using the HOS is that it consume more time for estimation than the PDS. The processing time for testing 100 images by the proposed method for different contaminations is tabulated in **Table 2**. The processing time results are obtained using an i5 Dell laptop running MATLAB software.

6. CONCLUSION

This work proposes a method for the automatic detection of weld defects in radiographic images. Previously, the Cepstral features were extracted from the Power Density Spectrum (PDS) using MUSIC method. We propose using Higher Order Spectra of the images (Bispectrum and Trispectrum) to extract the Cepstral features. The proposed method was tested and ran 8 times according to image degradations and image blurring. The degradations were Gaussian, Impulsive, Speckle and Poisson noise applied with and without image blurring. We conclude that in spite of time consumption, the proposed method yields best results for the automatic detection of weld defects in radiography images when the features were extracted from the Trispectrum of the image.

REFERENCES

1. J. T. Karlsen, G. Farrants, T. Torgrimsen and A. Reith, "Chemical Composition and Morphology of Welding Fume Particles and Grinding Dusts," American Industrial Hygiene Association Journal, Vol. 53, No. 5, 1992, pp. 290-297.
2. C. Hayes, "The ABC's of Nondestructive Weld Examination," Welding Journal, Vol. 76, No. 5, 1998, pp. 46-51.
3. W. K. F. Thai, "Development of a Computed Radiography-Based Weld Defect Detection and Classification System," M.Sc. Thesis, University of Sains, Penang, 2008.
4. H. R. Sim, M. B. Rabello, R. Camargo and M. S. Pereira, "Digital Radiography for the Inspection of Weld Seams of Pipelines—Better Sensitivity," Welding International, Vol. 24, No. 4, 2010, pp. 249-257.

5. G. Wang and T. W. Liao, "Automatic Identification of Different Types of Welding Defects in Radiographic Images," Nondestructive Testing and Evaluation: An Intertional Journal, Vol. 35, No. 8, 2002, pp. 519-528.

6. H. I. Shafeek, E. S. Gadelmawla, A. A. Abdel-Shafy and I. M. Elewa, "Automatic Inspection of Gas Pipeline Welding Defects Using an Expert Vision System," Nondestructive Testing and Evaluation: An International Journal, Vol. 37, No. 4, 2004, pp. 301-307.

7. T. Saravanan, S. Bagavathiappan, J. Philip, T. Jayakumar and B. Raj, "Segmentation of Defects from Radiographic Images by the Histogram Concavity Threshold Method," Insight, Vol. 49, No. 10, 2007, pp. 578-584.

8. S. S. L. Eong, Z. Samad, M. M. Ratnam and M. A. Khalid, "Weld Extraction from Digitized Radiographs Using Graphical Analysis of Weld Intensity Profiles," Journal of Technology, Vol. 45, No. D, 2006, pp. 167- 180.

9. J. Zapata, R. Vilar and R. Ruiz, "An Adaptive-NetworkBased Fuzzy Inference System for Classification of Welding Defects," Nondestructive Testing and Evaluation: An International Journal, Vol. 43, No. 3, 2010, pp. 191-199.

10. R. Vilar, J. Zapat and R. Ruiz, "Classification of Welding Defects in Radiographic Images Using an ANN with Modified Performance Function," Springer Berlin, Heidelberg, 2009, pp. 284-293.

11. J. Zapata, R. Vilar and R. Ruiz, "Automatic Inspection System of Welding Radiographic Images Based on ANN Under a Regularisation Process," Journal of Nondestructive Evaluation, Vol. 31, No. 1, 2012, pp. 34-45

12. H. Kasban, O. Zahran, H. Arafa, M. El-Kordy, S. M. S. Elaraby and F. E. A. El-Samie, "Welding Defect Detection from Radiographic Image Using Cepstral Approach," Nondestructive Testing and Evaluation: An International Journal, Vol. 44, No. 2, 2011, pp. 226-231.

13. H. Kasban, "Applying Advanced Digital Signal Processing Techniques in Industrial Radioisotopes Applications," Ph.D. Thesis, Faculty of Electronic Engineering, Menofia University, Shibin Al Kawm, 2012.

14. D. Coltuc, P. Bolon and J. M. Chassery, "Exact Histogram Specification," IEEE Transactions on Image Processing, Vol. 15, No. 5, 2006, pp.1143-1152.

15. I. Frosio and N. A. Borghese, "Statistical Based Impulsive Noise Removal in Digital Radiography," IEEE Transactions on Medical Imaging, Vol. 28, No. 1, 2009, pp. 3- 16.

16. T. W. Liao, D. M. Li and Y. M. Li, "Extraction of Welds from Radiographic Images Using Fuzzy Classifiers," Informatics and Computer Science: An International Journal, Vol. 126, No. 1-4, 2000, pp. 21-42.

17. M. A. Carrasco and D. Mery, "Segmentation of Welding Defects Using a Robust Algorithm," Materials Evaluation, Vol. 62, No. 11, 2004, pp. 1142-1147.

18. O. Zahran, H. Kasban, F. E. A. El-Samie and M. ElKordy, "Power Density Spectrum for the Identification of Residence Time Distribution Signals," Applied Radiation and Isotopes, Vol. 70, No. 11, 2012, pp. 2638-2645.

19. W. B. Collis, P. R. White and J. K. Hammond, "HigherOrder Spectra: The Bispectrum and Trispectrum," Mechanical Systems and Signal Processing, Vol. 12, No. 3, 1998, pp. 375-394.

20. M. G. Kang, K.-T. Lay and A. K. Katsaggelos, "Phase Estimation Using the Bispectrum and Its Application to Image Restoration," Optical Engineering, Vol. 30, No. 7, 1991, pp. 976-985.

21. J. F. McAloon, "Comparison of Higher Order Moment Spectrum Estimation Techniques," M.Sc. Thesis, Electrical Engineering, University of South Florida, Tampa, 1983.

22. N. Nafaâ, D. Redouane and B. Amar, "Weld Defect Extraction and Classification in Radiographic Testing Based Artificial Neural Networks," 15th WCNDT, Roma, 2000. http://www.ndt.net/article/wcndt00/papers/idn575/idn575.htm

23. E. S. Amin, "Application of Artificial Neural Networks to Evaluate Weld Defects of Nuclear Components," Journal of Nuclear and Radiation Physics, Vol. 3, No. 2, 2008, pp. 83-92.

24. C. M. Bishop, "Neural Networks for Pattern Recognition", Clarendon Press, Oxford, 1995.

25. J. Starck, F. Murtagh and A. Bijaoui, "Image Processing and Data Analysis: The Multiscale Approach," Cambridge University Press, Cambridge, 1998.

26. P. Gravel, G. Beaudoin and J. A. De Guise, "A Method for Modeling Noise in Medical Images," IEEE Transactions on Medical Imaging, Vol. 23, No. 10, 2004, pp. 1221-1232.

27. J. Harikiran, B. Saichandana and B. Divakar, "Impulse Noise Removal in Digital Images," International Journal of Computer Applications, Vol. 10, No. 8, 2010, pp. 39- 42.

28. P. R. Deshmukh and M. V. Sarode, "Reduction of Speckle Noise and Image Enhancement of Images Using Filtering Technique," International Journal of Advancements in Technology, Vol. 2, No. 1, 2011, pp. 30-38.

29. R. J. Ferrari and R. Winsor, "Digital Radiographic Image Denoising via Wavelet-Based Hidden Markov Model Estimation," Journal of Digital Imaging, Vol. 18, No. 2, 2005, pp. 154-167.

Chapter 5

Statistical Methods of SNP Data Analysis and Applications

Alexander Balatskiy[2], Larisa Samokhodskaya[2], Vsevolod Tkachuk[2]

[1]*Faculty of Mathematics and Mechanics, Moscow State University, Moscow, Russia*

[2]*Faculty of Basic Medicine, Moscow State University, Moscow, Russia*

ABSTRACT

We develop various statistical methods important for multidimensional genetic data analysis. Theorems justifying application of these methods are established. We concentrate on the multifactor dimensionality reduction, logic regression, random forests, stochastic gradient boosting along with their new modifications. We use complementary approaches to study the risk of complex diseases such as cardiovascular ones. The roles of certain combinations of single nucleotide polymorphisms and non-genetic risk factors are examined. To perform the data analysis concerning the coronary heart disease and myocardial infarction the Lomonosov Moscow State University supercomputer "Chebyshev" was employed.

Keywords: Genetic Data Statistical Analysis; Multifactor Dimensionality Reduction; Ternary Logic Regression; Random Forests; Stochastic Gradient Boosting; Independent Rule; Single Nucleotide Polymorphisms; Coronary Heart Disease; Myocardial Infarction

1. INTRODUCTION

In the last decade new high-dimensional statistical methods were developed for the data analysis (see, e.g., [1]). Special attention was paid to the study of genetic models (see, e.g., [2-4]). The detection of genetic susceptibility to complex diseases (such as diabetes and others) has recently drawn much attention in leading research centers. It is well-known that such diseases can be provoked by variations in different parts of the DNA code which are responsible for the formation of certain types of proteins. One of the most common individual's DNA variations is a single nucleotide polymorphism (SNP), i.e. a nucleotide change in a certain fragment of genetic code (for some percentage of population). Quite a number of recent studies (see, e.g., [5,6] and references therein) support the paradigm that certain combinations of SNP can increase the complex disease risk whereas separate changes may have no dangerous effect.

There are two closely connected research directions in genomic statistics. The first one is aimed at the disease risk estimation assuming the genetic portrait of a person is known (in turn this problem involves estimation of disease probability and classification of genetic data into high and low risk domains). The second trend is to identify relevant combinations of SNPs having the most significant influence, either pathogenic or protective.

In this paper we propose several new versions of statistical methods to analyze multidimensional genetic data, following the above-mentioned research directions. The methods developed generalize the multifactor dimensionality reduction (MDR) and logic regression (LR). We employ also some popular machine learning methods (see, e.g., [2]) such as random forests (RF) and stochastic gradient boosting (SGB).

Ritchie et al. [7] introduced MDR as a new method of analyzing gene-gene and gene-environment interactions. Rather soon the method became very popular. According to [8], since the first publication more than 200 papers applying MDR in genetic studies were written.

LR was proposed by Ruczinski et al. in [9]. Further generalizations are given in [6,10] and other works. LR is based on the classical binary logistic regression and the exhaustive search for relevant predictor combinations. For genetic analysis it is convenient to use explanatory variables taking 3 values. Thus we employ ternary variables and ternary logic regression (TLR), whereas the authors of the above-mentioned papers employ binary ones.

RF and SGB were initiated by Breiman [11] and Friedman [12] respectively. They belong to ensemble methods which combine multiple predictions from a certain base algorithm to obtain better predictive power. RF and SGB were successfully applied to genetics data in a number of papers (see [2,13] and references therein).

We compare various approaches on the real datasets concerning coronary heart disease (CHD) and myocardial infarction (MI). Each approach (MDR, TLR and machine learning) is characterized by its own way of constructing disease prediction algorithms. For each method one or several prediction algorithms admitting the least estimated prediction error are found (a typical situation is that there are several ones with almost the same estimated prediction error). These prediction algorithms provide a way to determine the domains where the disease risk is high or low. It is also possible to select combinations of SNPs and non-genetic risk factors influencing the liability to disease essentially. Some methods allow to present such combinations immediately. Other ones, which employ more complicated forms of dependence between explanatory and response variables, need further analysis based on modifications of permutation tests. New software implementing the mentioned statistical methods has been designed and used.

This work was started in 2010 in the framework of the general MSU project headed by Professors V. A. Sadovnichy and V. A. Tkachuk (see [14]). MSU supercomputer "Chebyshev" was employed to perform data analysis.

The rest of the paper is organized as follows. In Section 2 we discuss various statistical methods and prove theorems justifying their applications. Section 3 is devoted to analysis of CHD and MI datasets. Section 4 contains conclusions and final remarks.

2. METHODS

We start with some notation. Let N be the number of patients in the sample and let the vector $X^j = (X_1^j, \cdots, X_n^j)$ consist of genetic (SNP) and nongenetic risk factors of individual j $(j = 1, \cdots, N)$. Here n is the total number of factors and X_i^j is the value of the i-th variable (genetic or non-genetic factor) of individual j.

These variables are called explanatory variables or predictors. If X_i^j stands for a genetic factor (characterizes the i-th SNP of individual j) we set

$$X_i^j = \begin{cases} 0, & \text{SNP is homozygous for dominant allele,} \\ 1, & \text{SNP is heterozygous} \\ 2, & \text{SNP is homozygous for recessive allele.} \end{cases}$$

For biological background we refer, e.g., to [15].

We assume that non-genetic risk factors also take no more than three values, denoted by 0, 1 and 2. For example, we can specify a presence or absence of obesity (or hypercholesterolemia etc.) by the values 1 and 0 respectively. If a non-genetic factor takes more values (e.g., blood pressure), we can divide individuals into three groups according to its values.

Further on $X_1^j,\cdots,X_m^j$ stand for genetic data and $X_{m+1}^j,\cdots,X_n^j$ for non-genetic risk factors. Let a binary variable Y^j (response variable) be equal to 1 for a case, i.e. whenever individual j is diseased, and to −1 otherwise (for a control). Set

$$\xi=\left(\xi^1,\cdots,\xi^N\right)$$

where

$$\xi^j=(X^j,Y^j),\ \ j=1,\cdots,N\ .$$

Suppose $\xi^1,\cdots,\xi^N$ are i.i.d. random vectors. Introduce a random vector (X,Y) independent of ξ and having the same law as ξ^1. All random vectors (and random variables) are considered on a probability space $(\Omega,\mathcal{F},\mathrm{P})$, E denotes the integration w.r.t. P.

The main problem is to find a function in genetic and non-genetic risk factors describing the phenotype (that is the individual being healthy or sick) in the best way.

2.1. Prediction Algorithms

Let $\mathbb{X}:=\{0,1,2\}^n$ denote the space of all possible values of explanatory variables. Any function $f:\mathbb{X}\to\{-1,1\}$ is called a theoretical prediction function. Define the balanced or normalized prediction error for a theoretical prediction function f as

$$Err(f):=\mathrm{E}\left|Y-f(X)\right|\psi(Y)$$

where the penalty function $\psi:\{-1,1\}\to\mathbf{R}_+$. Obviously

$$\begin{aligned}Err(f)=&2\psi(-1)\mathrm{P}\left(f(X)=1,Y=-1\right)\\&+2\psi(1)\mathrm{P}\left(f(X)=-1,Y=1\right).\end{aligned} \quad (1)$$

Clearly $Err(f)$ depends also on the law of (X,Y) but we simplify the notation. Following [8,16] we put

$$\psi(y) = \frac{1}{4\mathrm{P}(Y = y)}, \quad y \in \{-1,1\},$$

where the trivial cases $\mathrm{P}(Y=-1)=0$ and $\mathrm{P}(Y=1)=0$ are excluded. Then

$$Err(f) = \frac{1}{2}\mathrm{P}\big(f(X)=1 \big| Y=-1\big) + \frac{1}{2}\mathrm{P}(f(X)=-1 \big| Y=1). \tag{2}$$

If $\mathrm{P}(Y=-1)=\mathrm{P}(Y=1)=1/2$ a sample is called balanced and one has $Err(f)=\mathrm{E}\left|Y-f(X)\right|/2$. Therefore in this case $Err(f)$ equals the classification error $\mathrm{P}(Y \neq f(X))$. In general,

$$Err(f) = \frac{1}{2}\mathrm{E}\left|Y^* - f(X^*)\right|.$$

with (X^*,Y^*) having the distribution

$$\mathrm{P}(X^*=x, Y^*=y) = \frac{1}{2}\mathrm{P}(X=x \big| Y=y),$$

$$(x,y) \in \mathcal{X} \times \{-1,1\}$$

The reason to consider this weighted scheme is that a misclassification in a more rare class should be taken into account with a greater weight. Otherwise, if the probability of disease $\mathrm{P}(Y=1)$ is small, then the trivial function $f(x) \equiv -1$ may have the least prediction error.

It is easy to prove that the optimal theoretical predicttion function minimizing the balanced prediction error is given by

$$f^*(x) = \begin{cases} 1, & p(x) > \mathrm{P}(Y=1), \\ -1, & \text{otherwise,} \end{cases} \tag{3}$$

Where

$$p(x) = \mathrm{P}(Y=1 \big| X=x), \quad x \in \mathcal{X}. \tag{4}$$

Then each multilocus genotype (with added non-genetic risk factors) $x \in \mathcal{X}$ is classified as high-risk if $f^*(x)=1$ or low-risk if $f^*(x)=-1$.

Since $p(x)$ and $\mathrm{P}(Y=1)$ are unknown, the immediate application of (3) is not possible. Thus we try to find an approximation of unknown function f^* using a prediction algorithm that is a function

$$f_{PA}=f_{PA}(x,\xi(S))$$

with values in $\{-1,1\}$ which depends on $x \in \mathcal{X}$ and the sample

$$\xi(S)=\{\xi^j, j \in S\}$$

Where

$$S \subset \{1,\cdots,N\}. \tag{5}$$

The simplest way is to employ formula (3) with $p(x)$ and $\mathrm{P}(Y=1)$ replaced by their statistical estimates. Consider

$$\hat{p}(x,\xi(S))=\frac{\sum_{j\in S} I\{Y^j=1, X^j=x\}}{\sum_{j\in S} I\{X^j=x\}}, \; x \in \mathcal{X}, \tag{6}$$

and take

$$\hat{\mathrm{P}}_S(Y=1)=\frac{1}{\#S}\sum_{j\in S} I\{Y^j=1\} \tag{7}$$

where $I\{A\}$ stands for the indicator of an event A and #D denotes the cardinality of a finite set D.

Along with (7) we consider

$$\hat{\mathrm{P}}_S(Y=1 \mid X \in C)=\frac{\sum_{j\in S} I\{Y^j=1, X^j \in C\}}{\sum_{j\in S} I\{X^j \in C\}}, \tag{8}$$

for $C \subset X$. Thus (6) is a special case of (8) for $C=\{x\}$ with $x \in X$. Note that a more difficult way is to search for the estimators of f^* using several subsamples of ξ.

For a given prediction algorithm f_{PA}, keeping in mind (2), we set

$$Err\left(f_{PA}(\cdot,\xi(S))\right) = \frac{1}{2}\sum\nolimits_{y\in\{-1,1\}} \mathrm{P}\left(f_{PA}(X,\xi(S)) \neq y \middle| Y=y\right). \quad (9)$$

If one deals with too many parameters, overfitting is likely to happen, i.e. estimated parameters depend too much on the given sample. As a result the constructed estimates give poor prediction on new data. On the other hand, application of too simple model may not capture the studied dependence structure of various factors efficiently. However the trade-off between the model's complexity and its predictive power allows to perform reliable statistical inference via new model validation techniques (see, e.g., [17]). The main tool of model selection is the cross-validation, see, e.g., [16]. Its idea is to estimate parameters by involving only a part of the sample (training sample) and afterwards use the remaining observations (test sample) to test the predictive power of the obtained estimates. Then an average over several realizations of randomly chosen training and test samples is taken, see [18].

As the law of (X,Y) is unknown, one can only construct an estimate $\widehat{Err}\left(f_{PA}(\cdot,\xi(S))\right)$ of $Err\left(f_{PA}(\cdot,\xi(S))\right)$. In Section 3 we use the estimated prediction error (EPE) of a prediction algorithm f_{PA} which is based on K-fold cross-validation $(K>1)$ and has the form

$$\widehat{Err}_K\left(f_{PA}(\cdot,\xi),\xi\right) = \frac{1}{2}\sum_{y\in\{-1,1\}}\frac{1}{K}\sum_{k=1}^{K}\frac{\sum_{(k)} I\left\{f_{PA}\left(X^j,\xi\left(\overline{S_k}\right)\right)\neq y, Y^j=y\right\}}{\sum_{(k)} I\left\{Y^j=y\right\}} \quad (10)$$

where the sum $\sum_{(k)}$ is taken over j belonging to

$$S_k = \{(k-1)[N/K]+1,\cdots, \\ k[N/K]I\{k<K\}+NI\{k=K\}\}, \tag{11}$$

$\overline{S_k} = \{1,\cdots,N\}\setminus S_k$ and $[a]$ is the integer part of $a\in\mathbf{R}$.

Let $\{k_1,\cdots,k_r\}\subset\{1,\cdots,n\}$. Introduce

$$C_{k_1,\cdots,k_r}(x)=\{u=(u_1,\cdots,u_n)\in \mathcal{X}: u_{k_i}=x_{k_i},\ i=1,\cdots,r\}.$$

The next result provides wide sufficient conditions for consistency of estimate (10).

Theorem 1. Suppose there exist a subset $U\subset\mathcal{X}$ and a subset $\{k_1,\cdots,k_r\}\subset\{1,\cdots,n\}$ such that the following holds:

1) For each $x\in\mathcal{X}$ and any finite dimensional vector v with components belonging to $\mathcal{X}\times\{-1,1\}$, functions $f_{PA}(\cdot,v)$ and f are constant on $C_{k_1,\cdots,k_r}(x)$.
2) For each $x\in U$ and any $W_N\subset\{1,\cdots,N\}$ with $\#W_N\to\infty$, one has $f_{PA}(x,\xi(W_N))\to f(x)$ a.s. when $N\to\infty$.
3) $\mathrm{P}(Y=1|X_{k_1}=x_{k_1},\ldots,X_{k_r}=x_{k_r})=\mathrm{P}(Y=1)$ if $x\in\mathcal{X}\setminus U.$
4) f is constant on $\mathcal{X}\setminus U$.

Then $\hat{Err}_K(f_{PA}(\cdot,\xi),\xi)\to Err(f)$ a.s., $N\to\infty$.

Remark. If we replace condition 3 of Theorem 1 by a more restrictive assumption 3') $\mathrm{P}(Y=1|X=x)=\mathrm{P}(Y=1)$ for all $x\in\mathcal{X}\setminus U$. then we can take $\{k_1,\cdots,k_r\}=\{1,\cdots,n\}$ to remove condition 1.

Proof. The proof is based on the following Lemma. Let $\{(Z_j^{(m)},Y_j^{(m)}),\ 1\le j\le m, m\in N\}$ be an array of rowwise independent random elements distributed as (Z,Y), where Z takes values in a finite set $\mathcal{Z}$ and Y takes values in $\{-1,1\}$. Assume that

$\{f_m(z),\ m \in N,\ z \in \mathcal{Z}\}$ is an array of random variables with values in $\{-1,1\}$.

Suppose there exists $U \subset \mathcal{Z}$ such that the following conditions hold:

1) $f_m(z) \to f(z)$ a.s. for all $z \in U$, as $m \to \infty$, where a nonrandom function $f : \mathcal{Z} \to \{-1,1\}$.

2) $\mathrm{P}(Y=1|Z=z) = \mathrm{P}(Y=1)$ if $z \in \mathcal{Z} \setminus U.$

3) f is constant on $\mathcal{Z} \setminus U.$

Then, as $m \to \infty,$

$$\frac{1}{2}\sum_{y\in\{-1,1\}} \frac{\sum_{j=1}^{m} I\left\{f_m\left(Z_j^{(m)}\right) \neq y, Y_j^{(m)} = y\right\}}{\sum_{j=1}^{m} I\left\{Y_j^{(m)} = y\right\}} \to Err(f) \ \text{a.s.} \tag{12}$$

Proof of Lemma. Set $Q_m(y) = \sum_{j=1}^{m} I\{Y_j^{(m)} = y\}$ and define events

$$A_j^{(m)}(y) = \left\{f_m\left(Z_j^{(m)}\right) \neq y, Y_j^{(m)} = y\right\},$$

$$B_j^{(m)}(y) = \left\{f(Z_j^{(m)}) \neq y, Y_j^{(m)} = y\right\}.$$

Then the l.h.s. of (12) equals

$$\sum_{y\in\{-1,1\}} \frac{1}{2Q_m(y)} \sum_{j=1}^{m} I\left\{A_j^{(m)}(y)\right\}.$$

For $y \in \{-1,1\}$, we have

$$\frac{1}{Q_m(y)}\sum_{j=1}^{m} I\left\{A_j^{(m)}(y)\right\}$$
$$=\frac{1}{m\mathrm{P}\{Y=y\}}\sum_{j=1}^{m} I\left\{A_j^{(m)}(y)\right\}$$
$$+\frac{1}{m}\sum_{j=1}^{m} I\left\{A_j^{(m)}(y)\right\}\left(\frac{m}{Q_m(y)}-\frac{1}{\mathrm{P}(Y=y)}\right). \tag{13}$$

The absolute value of the second term in the r.h.s. of (13) does not exceed $\left|m/(Q_m(y))-1/\mathrm{P}(Y=y)\right|$. and tends to 0 a.s. if $m\to\infty$. This statement follows from the strong law of large numbers for arrays (SLLNA), see [19].

Note that

$$\frac{1}{m}\sum_{j=1}^{m} I\left\{A_j^{(m)}(y)\right\}$$
$$=\frac{1}{m}\sum_{j=1}^{m} I\left\{B_j^{(m)}(y)\right\}$$
$$+\frac{1}{m}\sum_{j=1}^{m} I\left\{Z_j^{(m)}\in U\right\}\left(I\left\{A_j^{(m)}(y)\right\}-I\left\{B_j^{(m)}(y)\right\}\right)$$
$$+\frac{1}{m}\sum_{j=1}^{m} I\left\{Z_j^{(m)}\notin U\right\}\left(I\left\{A_j^{(m)}(y)\right\}-I\left\{B_j^{(m)}(y)\right\}\right). \tag{14}$$

According to SLLNA the first term in the r.h.s. of (14) goes to $P(f(Z)\neq y, Y=y)$ a.s. We claim that the second term tends to 0 a.s. Indeed, the set Z is finite and the functions f_m f, take only two values. Therefore, by condition 1, for almost all $\omega\in\Omega$, there exists $N_1=N_1(\omega)$ such that $f_m(z)=f(z)$ for all $z\in U$ and $m>N_1$. Hence second term in the r.h.s. of (14) equals 0 for all $m>N_1$, which proves the claim. Thus, it remains to estimate the third term.

In view of condition 3, w.l.g. we may assume that $f(z)=-1$ for $z\in Z\setminus U$. Then we obtain

$$
\begin{aligned}
V_m &:= \sum_{y\in\{-1,1\}} \sum_{j=1}^{m} \frac{I\{Z_j^{(m)} \notin U\}\left(I\{A_j^{(m)}(y)\} - I\{B_j^{(m)}(y)\}\right)}{m\mathrm{P}(Y=y)} \\
&= \frac{1}{m}\sum_{j=1}^{m} I\{Z_j^{(m)} \notin U\} I\{f_m(Z_j^{(m)}) = 1\} R_j^{(m)} \\
&= \frac{1}{m}\sum_{z\in Z\setminus U} I\{f_m(z) = 1\} \sum_{j=1}^{m} I\{Z_j^{(m)} = z\} R_j^{(m)}
\end{aligned} \tag{15}
$$

where

$$
R_j^{(m)} = \frac{I\{Y_j^{(m)} = -1\}}{\mathrm{P}(Y=-1)} - \frac{I\{Y_j^{(m)} = 1\}}{\mathrm{P}(Y=1)}.
$$

SLLNA and condition 2 imply that, for $z \in Z\setminus U$ and $y \in \{-1,1\}$, we have

$$
\begin{aligned}
&\sum_{j=1}^{m} \frac{I\{Z_j^{(m)} = z\} I\{Y_j^{(m)} = y\}}{m\mathrm{P}(Y=y)} \\
&\to \frac{\mathrm{P}(Z=z, Y=y)}{\mathrm{P}(Y=y)} = \mathrm{P}(Z=z)
\end{aligned}
$$

almost surely. Therefore, for almost all $\omega \in \Omega$, there exists $N_2 = N_2(\omega)$ such that

$$
\left| \frac{1}{m}\sum_{j=1}^{m} I\{Z_j^{(m)} = z\} R_j^{(m)} \right| < \varepsilon
$$

for all $z \in Z\setminus U$ and $m > N_2$. Using the last estimate and (15) we finally get that, for $m > N_2$,

$$
|V_m| \le \sum_{z\in Z\setminus U} \varepsilon\, I\{f_m(z) = 1\} \le \varepsilon \cdot \#Z.
$$

Hence $V_m \to 0$ a.s. if $m \to \infty$. Combining (12)-(16) we obtain the desired result. □

Let us return to the proof of Theorem 1. Fix $1 \le k \le K$ and take

$$f_m(z) := f_{PA}\left(x, \xi\left(\overline{S_k}\right)\right)$$

where $z \in Z := \{0,1,2\}^r$, $m := \# S_k$, S_k is introduced in (11), and x is any element of X with $(x_{k_1}, \cdots, x_{k_r}) = z.$ By condition 1 of Theorem 1, f_m is well defined.

Applying Lemma to arrays

$$\left\{\left(Z_j^{(m)}, Y_j^{(m)}\right),\ 1 \le j \le m\right\} := \left(\left(X_{k_1}^j, \ldots, X_{k_r}^j\right), Y^j\right), j \in S_k\right\}$$

and $\{f_m(z),\ z \in Z\}$, we obtain that almost surely

$$\frac{1}{2}\sum_{y \in \{-1,1\}} \frac{\Sigma_{(k)} I\left\{f_{PA}\left(X^j, \xi\left(\overline{S_k}\right)\right) \ne y, Y^j = y\right\}}{\Sigma_{(k)} I\left\{Y^j = y\right\}} \to Err(f)$$

as $\# S_k \to \infty$. Thus

$$\widehat{Err}_K\left(f_{PA}(\cdot, \xi), \xi\right) \to \frac{1}{K}\sum_{k=1}^{K} Err(f) = Err(f)$$

when $N \to \infty$, which completes the proof of Theorem 1.

An important problem is to make sure that the predicttion algorithm f_{PA} gives statistically reliable results. The quality of an algorithm is determined by its predicttion error (9) which is unknown and therefore the inference is based on consistent estimates of this error. Clearly the high quality of an algorithm means that it captures the dependence between predictors and response variables, so the error is made more rarely than it would be if these variables were independent. Consider a null hypothesis H_0 that X and Y are independent. If they are in fact dependent, then for any reasonable prediction algorithm f_{PA} an appropriate test procedure involving f_{PA} should reject H_0 at the significance level α, e.g., 5%. This shows that the results of the algorithm could not be obtained by chance. For such a procedure, we take a permutation test which allows to find the Monte Carlo estimate $\widehat{p}=\widehat{F}\left(\widehat{Err}_K\left(f_{PA},\xi\right)\right)$ (see [20]) of the true p-value $p=F\left(\widehat{Err}_K\left(f_{PA},\xi\right)\right)$.

Here F is the cumulative distribution function (c.d.f.) of $\widehat{Err}_K$ under H_0 and $\widehat{F}=\widehat{F}(z)$ is the corresponding empirical c.d.f. We reject H_0 if $\widehat{p}<\alpha$. For details we refer to [21].

Now we pass to the description of various statistical methods and their applications (in Section 3) to the cardiovascular risk detection.

2.2. Multifactor Dimensionality Reduction

MDR is a flexible non-parametric method of analyzing gene-gene and gene-environment interactions. MDR does not depend on a particular inheritance model. We give a rigorous description of the method following ideas of [7] and [8].

To calculate the balanced error of a theoretical predicttion function we use formula (2). Note that the approach based on penalty functions is not the only possible. Nevertheless it outperforms substantially other approaches involving overand undersampling (see [8]).

As mentioned earlier, the probability p(x) introduced in (4) is unknown. To find its estimate one can apply maximum likelihood approach assuming that the random variable $I\{Y=1\}$ conditionally on $X=x$ has a Bernoulli distribution with unknown parameter $p(x)$. Then we come to (6).

A direct calculation of estimate (6) with exhaustive search over all possible values of x is highly inefficient, since the number of different values of x grows exponentially with number of risk factors. Moreover, such a search leads to overfitting. Instead, it is usually supposed that $p(x)$ depends non-trivially not on all, but on certain variables x_i. That is, there exist $l\in\mathbf{N},\ l<n,$ and a

vector $(k_1^*,\cdots,k_l^*)$, where $1 \le k_1^* < \cdots < k_l^* \le n$, such that for each $x=(x_1,\cdots,x_n)\in \mathcal{X}$, the following relation holds:

$$p(x)=\mathrm{P}\left(Y=1 \middle| X_{k_1^*}=x_{k_1^*},\cdots,X_{k_l^*}=x_{k_l^*}\right) \quad .(17)$$

In other words only few factors influence the disease and other ones can be neglected. A combination of indices $(k_1^*,\cdots,k_l^*)$, in formula (17) having minimal l is called the most significant.

For $x\in \mathcal{X}$ and indices $\{k_1,\cdots,k_r\}\subset\{1,\cdots,n\}$, set

$$f_{k_1,\cdots,k_r}(x) = \begin{cases} 1, & \mathrm{P}(Y=1 \mid X_{k_1}=x_{k_1},\dots,X_{k_r}=x_{k_r}) > \mathrm{P}(Y=1), \\ -1, & \text{otherwise.} \end{cases}$$

Consider the estimate

$$\hat{f}_{k_1,\cdots,k_r}(x,\xi(S)) := \begin{cases} 1, & \hat{\mathrm{P}}_S(Y=1 \mid X\in C_{k_1,\cdots,k_r}(x)) > \hat{\mathrm{P}}_S(Y=1), \\ -1, & \text{otherwise,} \end{cases} \quad (18)$$

where S is introduced in (5).

Theorem 2. Let $(k_1^*,\cdots,k_l^*)$ be the most significant combination. Then for any fixed $\{k_1,\cdots,k_r\}\subset\{1,\cdots,n\}$ one has

1) $Err\left(f_{k_1^*,\dots,k_l^*}\right) \le Err\left(f_{k_1,\cdots,k_r}\right)$;

2) $\widehat{Err}_K\left(\hat{f}_{k_1,\cdots,k_r}\right)$ is a strongly consistent asymptotically unbiased estimate of $Err\left(f_{k_1,\cdots,k_r}\right)$ as $N\to\infty$;

3) for any $\varepsilon,\ \delta > 0$ and all N large enough

$$\mathrm{P}\left(\widehat{Err}_K\left(\widehat{f}_{k_1^*,\cdots k_l^*}\right) < \widehat{Err}_K\left(\widehat{f}_{k_1,\cdots,k_r}\right) + \varepsilon\right) > 1 - \delta$$

Proof. 1) It follows from (17) that $f_{k_1^*,\cdots,k_l^*}$ coinsides with function f^* (see (3)) which has the minimal balanced prediction error.

2) Let us verify the conditions of Theorem 1 for $f_{PA}(x,\xi(S)) := \widehat{f}_{k_1,\cdots,k_r}(x,\xi(S))$. Condition 1 follows from the definition of $\widehat{f}_{k_1,\cdots,k_r}(x,\xi(S))$. Further, put

$$U := \left\{x \in X : \mathrm{P}\left(Y = 1 \middle| C_{k_1,\cdots,k_r}(x)\right) \neq \mathrm{P}(Y = 1)\right\} \tag{19}$$

where $C_{k_1,\cdots,k_r}$ was introduced in Section 2.1 before Theorem 1. We claim that, for each $x \in U$ and any $W_N \subset \{1,\cdots,N\}$ such that $\#W_N \to \infty$, the following relation holds:

$$\widehat{f}_{k_1,\cdots,k_r}(x,\xi(W_N)) \to f_{k_1,\cdots,k_r}(x) \text{ a.s. if } N \to \infty.$$

Indeed, assume that, for some $\varepsilon > 0,$

$$\mathrm{P}\left(Y = 1 \middle| X_{k_1} = x_{k_1},\cdots,X_{k_r} = x_{k_r}\right) - \mathrm{P}(Y = 1) > \varepsilon.$$

SLLNA implies that

$$\widehat{\mathrm{P}}_{W_N}\left(Y = 1 \middle| X \in C_{k_1,\cdots,k_r}(x)\right) - \widehat{\mathrm{P}}_{W_N}(Y = 1)$$

converges a.s. to

$$\mathrm{P}\left(Y = 1 \middle| X_{k_1} = x_{k_1},\cdots,X_{k_r} = x_{k_r}\right) - \mathrm{P}(Y = 1),$$

$N \to \infty$. Then, for almost all $\omega \in \Omega,$ there exists $N_0 = N_0(\omega)$ such that

$$\hat{\mathrm{P}}_{W_N}\left(Y=1 \mid X \in C_{k_1,\dots,k_r}(x)\right)-\hat{\mathrm{P}}_{W_N}\left(Y=1\right)>\varepsilon/2$$

for all $N>N_0$. Therefore, for all $N>N_0$, we have $\hat{f}_{k_1,\dots,k_r}\left(x,\xi(W_N)\right)=1=f_{k_1,\dots,k_r}(x)$, which proves the claim. Thus condition 2 of Theorem 1 is met.

Conditions 3 and 4 of Theorem 1 follow from (19) and the definitions of $\hat{f}_{k_1,\dots,k_r}\left(x,\xi(S)\right)$ and $f_{k_1,\dots,k_r}(x)$.

Since all conditions of Theorem 1 are satisfied, we have $\widehat{Err}_K\left(\hat{f}_{k_1,\dots,k_r}\right)\to Err\left(f_{k_1,\dots,k_r}\right)$ a.s. and in mean (due to the Lebesgue theorem as $\widehat{Err}_K\left(\hat{f}_{k_1,\dots,k_r}\right)$ is bounded by 1).

3) Follows from 1) and 2).

In view of this result it is natural to pick one or a few combinations of factors with the smallest EPEs as an approximation for the most significant combination.

The last step in MDR is to determine statistical significance of the results. Here we test a null hypothesis of independence between predictors X and response variable Y. This can be done via the permutation test mentioned in Section 2.1.

MDR method with "independent rule". We propose multifactor dimensionality reduction with "independent rule" (MDRIR) method to improve the estimate of probability $p(x)$. This approach is motivated by [22] which deals with classification of large arrays of binary data. The principal difficulty with employment of formula (6) is that the number of observations in numerator and denominator of the formula might be small even for large N (see, e.g., [23]). This can lead to inaccurate estimates and finally to a wrong prediction algorithm. Moreover, for some samples the denominator of (6) might equal zero.

The Bayes formula implies that $p(x)$ equals

$$\frac{\mathrm{P}\left(X=x \mid Y=1\right)\mathrm{P}\left(Y=1\right)}{\mathrm{P}\left(X=x \mid Y=1\right)\mathrm{P}\left(Y=1\right)+\mathrm{P}\left(X=x \mid Y=-1\right)\mathrm{P}\left(Y=-1\right)} \quad (20)$$

where the trivial cases $\mathrm{P}(Y=-1)=0$ and $\mathrm{P}(Y=1)=0$ are excluded. Substituting (20) into (3) we obtain the following expression for the prediction function:

$$f^{*}(x)=\begin{cases} 1, & \mathrm{P}(X=x\,|\,Y=1)>\mathrm{P}(X=x\,|\,Y=-1), \\ -1, & \text{otherwise.} \end{cases} \tag{21}$$

As in standard MDR method described above, we assume that formula (17) holds. It was proved in [22] that for a broad class of models (e.g., Bahadur and logit models) the conditional probability

$$\mathrm{P}\left(X_{k_1}=x_1,\cdots,X_{k_r}=x_r \,\middle|\, Y=y\right)$$

where $y=\pm1,$ can be estimated in the following way:

$$\hat{\mathrm{P}}_S\left(X_{k_1}=x_1,\ldots,X_{k_r}=x_r \,\middle|\, Y=y\right) := \prod_{i=1}^{r}\hat{\mathrm{P}}_S\left(X_{k_i}=x_i \,\middle|\, Y=y\right), \tag{22}$$

here (cf. (8))

$$\hat{\mathrm{P}}_S\left(X_{k_i}=x \,\middle|\, Y=y\right)=\frac{\sum_{j\in S} I\left\{X_{k_i}^{j}=x, Y^{j}=y\right\}}{\sum_{j\in S} I\left\{Y^{j}=y\right\}}. \tag{23}$$

Combining (17) and (21)-(23) we find the desired estimate of $f^{*}(x)$.

A number of observations in numerator and denominator of (23) increases considerably comparing with (18). It allows to estimate the conditional probability more precisely whenever the estimate introduced in (22) is reasonable. For instance, sufficient conditions justifying the application of (22) are provided in [22, Cor.5.1]. MDRIR might have some advantage over MDR in case when the size l of the most significant combination $(k_1^*,\dots,k_l^*)$ is large. However, MDR for small l can demonstrate better behavior than MDRIR.

Thus, as opposed to standard MDR method, MDRIR uses alternative estimates of conditional probabilities. All other steps (prediction algorithm construction, EPE calculation) remain the same. As far as we know, this modification of MDR has not been applied before. It is based on a combination of the original MDR method (see [7]) and the ideas of [22].

2.3. Ternary Logic Regression

LR is a semiparametric method detecting the most significant combinations of predictors as well as estimating the conditional probability of the disease.

Let $p^*(x)=\mathrm{P}(Y^*=1|X^*=x)$ (where $x\in\mathbf{X}$) be the conditional probability of a disease defined in normalized sample, where (X^*,Y^*) was introduced in Section 2.1. Note that formula (3) can be written as follows:

$$f^*(x)=\begin{cases}1, & p^*(x)>1/2,\\ -1, & \text{otherwise.}\end{cases}$$

We suppose that trivial situations when $p^*(x)\in\{0,\ 1\}$ do not occur and omit them from the consideration. To estimate $p^*(x)$ we pass to the logistic transform

$$q^*(x)=\lambda(p^*(x)) \tag{24}$$

where $\lambda(z)=\log(z/(1-z)),\ z\in(0,1)$, is the inverse logistic function. The logistic function equals $\Lambda(t)=(1+e^{-t})^{-1}$, $t\in\mathbf{R}$. Note that we are going to estimate the unknown disease probability with the help of linear statistics with appropriately selected coefficients. Therefore it is natural to avoid restrictions on possible values of the function estimated. Thus the logistic transform is convenient, as $p^*(x)\in(0,\ 1)$ for $x\in\mathbf{X}$, while $q^*(x)$ can take all real values.

Consider a class $\mathcal{G}$ of all real-valued functions in ternary variables $x_1,\cdots,x_n$. We call a model of the dependence between the disease and explanatory variables any subclass $M\subset\mathcal{G}$. Set

$$\widehat{\psi}\left(y,\xi(S)\right)=\frac{1}{4\hat{\mathrm{P}}_S\left(Y=y\right)},\ y\in\{-1,1\},$$

with $\hat{\mathrm{P}}_S\left(Y=y\right)$ appearing in (7). Define the normalized smoothed score function

$$L\left(h,\xi(S)\right)=\frac{1}{\#S}\sum_{j\in S}\varphi\left(-Y^j h\left(X^j\right)\right)\hat{\psi}\left(Y^j,\xi(S)\right) \tag{25}$$

where S was introduced in (5), $\varphi(t)=\log_2\left(1+e^t\right)$ for $t\in\mathbf{R}$, and $h\in M$. In contrast to previous works our version of LR (more precisely, TLR) scheme involves normalization (cf. (1)), i.e. taking the observations with weights dependent on the proportion of cases and controls in subsample $\xi(S)$. An easy computation yields that $\arg\min_{h\in M}L\left(h,\xi(S)\right)$ equals $\arg\max_{h\in M}$ of the function

$$\frac{1}{\#S}\sum_{j\in S}\left(\frac{I\{Y^j=1\}}{2\hat{\mathrm{P}}_S\left(Y=1\right)}\log\gamma_j+\frac{I\{Y^j=-1\}}{2\hat{\mathrm{P}}_S\left(Y=-1\right)}\log\left(1-\gamma_j\right)\right)$$

with $\gamma_j=\Lambda\left(h\left(X^j\right)\right)$. That is, minimizing the score function is equivalent to the normalized maximum likelihood estimation of q^*.

The next theorem guarantees strong consistency of this estimation method whenever the model is correctly specified, i.e. $q^*\in M$. To formulate this result introduce $h\left(\cdot,\xi(S)\right):=\arg\min_{q\in M}L\left(q,\xi(S)\right)$.

Theorem 3. Let $h_0 \equiv 0$ $q^* \in M$, belong to M and

$$\min_{(x,y)\in\mathcal{X}\times\{-1,1\}} P(X=x|Y=y)>0.$$

Consider $W_N \subset \{1,\cdots,N\}$ and set

$$h_N(\cdot)=h(\cdot,\xi(W_N)).$$

Then $h_N(x)\to q^*(x)$ a.s. for all $x\in\mathcal{X}$ when $\#W_N\to\infty$. Moreover,

$Err_K(f_{PA}(\cdot,\xi),\xi)\to Err(f^*)$ a.s., $N\to\infty$

where $f_{PA}(\cdot,\xi)=2I\{\Lambda(h(\cdot,\xi))>1/2\}-1$.

Proof. At first we show that

$$h_N(x)<\varphi^{-1}\left(\frac{4}{\mathrm{P}(X=x|Y=-1)}\right) \text{ a.s.}$$

for any $x\in\mathcal{X}$ and all $N>N_1=N_1(\omega)$. Put $l_N = \#W_N$. By definition

$$\frac{\varphi(h_N(x))}{4l_N}\sum_{j\in W_N}\frac{I\{X^j=x,Y^j=-1\}}{\hat{\mathrm{P}}_{W_N}(Y=-1)}\le L(h_N,\xi(W_N))$$

$$\le L(0,\xi(W_N))=\frac{1}{4l_N}\sum_{(j,y)\in W_N\times\{-1,1\}}\frac{I\{Y^j=y\}}{\hat{\mathrm{P}}_{W_N}(Y=y)}.$$

Using SLLNA, we get that a.s.

$$\max_{x\in\mathcal{X}}\left|\frac{1}{l_N}\sum_{j\in W_N}\frac{I\{X^j=x,Y^j=-1\}}{\hat{\mathrm{P}}_{W_N}(Y=-1)}\right.$$

$$\left.-\mathrm{P}(X=x|Y=-1)\right|\to 0.$$

Obviously

$$\frac{1}{l_N}\sum_{(j,y)\in W_N\times\{-1,1\}}\frac{I\{Y^j=y\}}{\hat{P}_{W_N}(Y=y)}=2.$$

These relations imply the desired estimate of $h_N(x)$. Similarly we prove that

$$h_N(x)>-\varphi^{-1}\left(\frac{4}{P(X=x|Y=1)}\right)$$

for any $x\in\mathcal{X}$ and all $N>N_2=N_2(\omega)$. Consequently, we see that $h_N\in M_C:=M\cap\{h:\|h\|_\infty\le C\}$ for $N>\max(N_1,N_2)$,

here $\|h\|_\infty=\max_{x\in\mathcal{X}}|h(x)|$ and

$$C=\max_{(x,y)\in\mathcal{X}\times\{-1,1\}}\varphi^{-1}\left(\frac{4}{P(X=x|Y=y)}\right).$$

If $h\in M_C$ then $\left|L(h,\xi(W_N))-E\varphi(-Yh(X))\psi(Y)\right|$ is less than

$$\frac{\varphi(\|h\|_\infty)}{2}\sum_{y\in\{-1,1\}}\left(\left|\frac{1}{\hat{P}_{W_N}(Y=y)}-\frac{1}{P(Y=y)}\right|\right.$$
$$\left.+\sum_{x\in\mathcal{X}}\left|\sum_{j\in W_N}\frac{I\{X^j=x,Y^j=y\}}{l_N P(Y=y)}-\frac{P(X^j=x,Y^j=y)}{P(Y=y)}\right|\right).$$

By SLLNA, we get that

$$L(h,\xi(W_N))-E\varphi(-Yh(X))\psi(Y)\to 0 \quad \text{a.s.} \quad (26)$$

uniformly over $\{h:\|h\|_\infty\le C\}$. Note also that

$$2E\varphi(-Yh(X))\psi(Y)=E\varphi(-Y^*h(X^*))$$
$$=-E\log_2\left(\Lambda(h(X^*))^{I\{Y^*=1\}}\left(1-\Lambda(h(X^*))\right)^{1-I\{Y^*=1\}}\right).$$

By the (conditional) information inequality,

$$\mathrm{E}\log_2\left(\Lambda\left(h\left(X^*\right)\right)^{I\{Y^*=1\}}\left(1-\Lambda\left(h\left(X^*\right)\right)\right)^{1-I\{Y^*=1\}}\right).$$

attains its maximum over all functions h only at q^* introduced in (24). Under conditions of the theorem, $q^* \in M$. Therefore, by definitions of h_N and q^* we have

$$L\left(h_N,\xi\left(W_N\right)\right)\leq L\left(q^*,\xi\left(W_N\right)\right),$$

$$\mathrm{E}\varphi\left(-Y^*h\left(X^*\right)\right)|_{h=h_N}\geq \mathrm{E}\varphi\left(-Y^*q^*\left(X^*\right)\right).$$

By (26) and SLLNA

$$L\left(h_N,\xi\left(W_N\right)\right)-\frac{1}{2}\mathrm{E}\varphi\left(-Y^*h\left(X^*\right)\right)|_{h=h_N}\to 0 \text{ a.s.,}$$

$$L\left(q^*,\xi\left(W_N\right)\right)\to\frac{1}{2}\mathrm{E}\varphi\left(-Y^*q^*(X^*)\right) \text{ a.s.}$$

Hence

$$\mathrm{E}\varphi\left(-Y^*h\left(X^*\right)\right)|_{h=h_N}\to \mathrm{E}\varphi\left(-Y^*q^*\left(X^*\right)\right) \text{ a.s.}$$

This is possible only when $h_N\left(x\right)\to q^*\left(x\right)$ a.s. for all $x\in\mathcal{X}.$ Indeed, for almost all $\omega\in\Omega,$ we can always take a subsequence $h_{N_k}=h_{N_k(\omega)}\left(\cdot,\omega\right)$ converging to some function $\mu=\mu\left(\cdot,\omega\right)$ with
$\mathrm{E}\varphi\left(-Y^*\mu\left(X^*\right)\right)=\mathrm{E}\varphi\left(-Y^*q^*\left(X^*\right)\right).$

Hence, by the information inequality, $\mu(\cdot,\omega)=q(\cdot).$

To establish the second part of Theorem 3 we note that $f_{PA}\left(x,\xi\right)=2I\left\{\Lambda\left(h\left(x,\xi\right)\right)>1/2\right\}-1$ converges a.s. to $f^*\left(x\right)=2I\left\{\Lambda\left(q^*\left(x\right)\right)>1/2\right\}-1$ for all $x\in U$ where

$$U := \{x \in \mathcal{X} : p^*(x) \neq 1/2\}$$
$$= \{x \in \mathcal{X} : \mathrm{P}(Y=1|X=x) \neq \mathrm{P}(Y=1)\}.$$

Then the conclusion follows from Remark after Theorem 1. The proof is complete.

A wide and easy to handle class of models is obtained by taking functions linear in variables $x_1, \cdots, x_n$. or/and in their products. In turn these functions admit a convenient representation by elementary polynomials (EP). Recall that EP is a function T in ternary variables $x_1, \cdots, x_n$. belonging to $\{0,1,2\}$ which can be represented as a finite sum of products $x_1^{u_1} \cdots x_n^{u_n}$ where $u_1, \cdots, u_n \in \mathbf{Z}_+$. The addition and multiplication of ternary variables is considered by modulo 3. Any EP can be represented as a binary tree in which knots (vertices which are not leaves) contain either the addition or multiplication sign, and each leaf corresponds to a variable. Different trees may correspond to the same EP, thus this relation is not one-to-one. However, it does not influence our problem, so we keep the notation T for a tree. A finite set of trees $F = (T_1, \cdots, T_s)$ is called a forest. For a tree T its complexity C(T) is the number of leaves. The complexity C(F) of a forest F is the maximal complexity of trees constituting F. It is clear that if $g \in \mathcal{G}$ then there exists $s \geq 1$ such that g has the form

$$g(x_1, \cdots, x_n) = \beta_0 + \sum_{i=1}^{s} \beta_i T_i(x_1, \cdots, x_n), \tag{27}$$

here $\beta_0, \beta_1, \cdots, \beta_s \in \mathbf{R}$ and $T_1, \cdots, T_s$ are EP.

Let us say that function g belongs to a class $\mathcal{G}_r(s)$, where $s, r \in \mathbf{N}$, if there exist a decomposition (27) of g such that all trees T_i ($i = 1, \cdots, s$) have complexity less or equal to r. We identify a function $g \in \mathcal{G}_r(s)$ with pair (F, β) where F is the corresponding forest and $\beta = (\beta_0, \cdots, \beta_s)$ is the vector of coefficients in (27).

Minimization of $L(h,\xi(S))$ defined by (25) over all functions $h \in M \subset \mathcal{G}_r(s)$ is done in two alternating steps. First, we find the optimal value of b while F is fixed (which is the minimization of a smooth function in several variables) and then we search for the best F. The main difficulty is to organize this search efficiently. Here one uses stochastic algorithms, since the number of such forests increases rapidly when the complexity r grows. For $s \in \mathbf{N}$, a forest $F=(T_1,\cdots,T_s)$ and a subsample $\xi(S)$ (see (5)), consider a prediction algorithm f_{LR}^{F} setting

$$f_{\mathrm{LR}}^{F}(x)=\begin{cases} 1, & \hat{h}(x)>0, \\ -1, & \text{otherwise,} \end{cases}$$

where $\hat{h}=(F,\widehat{\beta})$ and

$$\widehat{\beta}=\arg\min_{\beta} L\left(\beta_0+\sum_{j=1}^{s}\beta_j T_j(\cdot),\xi(S)\right).$$

Define also the normalized prediction error of a forest $F=(T_1,\cdots,T_s)$ as $\phi(F)=\widehat{Err}_K\left(f_{\mathrm{LR}}^{F}(\cdot,\xi),\xi\right)$.

A subgraph B of a tree T is called a branch if it is itself a binary tree (i.e. it can be obtained by selecting one vertex of T together with its offspring). The addition and multiplication signs standing in a knot of a tree are called operations, thus * stands for sum or product. Following [9], call the tree T' a neighbor of T if it is obtained from T via one and only one of the following transformations.

1) Changing one variable to another in a leaf of the tree T (variable change).

2) Replacing an operation in a knot of a tree T with another one, i.e. sum to product or vice versa (operator change).

3) Changing a branch of two leaves to one of these leaves (deleting a leaf).

4) Changing a leaf to a branch of two leaves, one of which contains the same variable as in initial leaf (splitting a leaf).

5) Replacing a branch $B_1 * B_2$ with the branch B_1 (branch pruning).

6) Changing a branch B to a branch $x_i * B$ (branch growing), here x_i is a variable.

We say that forests F and F' are neighbors if they can be written as $F=(T_1,\cdots,T_s)$ and $F'=(T_1',\cdots,T_s)$ where T_1 and T_1' are neighbors. The neighborhood relation defines a finite connected graph on all forests of equal size s with complexity not exceeding r. To each vertex F of this graph we assign a number $\phi(F)$. To find the global minimum of a function defined on a finite graph we apply the simulated annealing method (see, e.g., [24]). This method constructs some specified Markov process which takes values in the graph vertices and converges with high probability to the global minimum of the function. To avoid stalling at a local minimal point the process is allowed to pass with some small probability to a point F having greater value of $\phi(F)$ than current one. We propose a new modification of this method in which the output is the forest corresponding to the minimal value of a function $\phi(F)$ over all (randomly) visited points. Since simulated annealing performed involves random walks on a complicated graph consisting of trees as vertices, the algorithm was realized by means of the MSU supercomputer.

2.4. Machine Learning Methods

Let us describe two machine learning methods: random forests and stochastic gradient boosting. They will be used in Section 3.

We employ classification and regression trees (CART) as a base learning algorithm in RF and SGB because it showed good performance in a number of studies (see [18]). Classification tree T is a binary tree having the following structure. Any leaf of T contains either 1 or –1 and for any vertex P in T (including leaves) there exists a subset A_P of the explanatory variable space X, such that the following properties hold:

1) $A_P = X$, if P is the root of T.

2) If vertices P' and P'' are children of P, then

$A_{P'} \cup A_{P''} = A_P$ and $A_{P'} \cap A_{P''} = \emptyset$.

In particular, subsets corresponding to the leaves form the partition of $\mathcal{X}$. A classifier defined by a classification tree is introduced as follows. To obtain a prediction of Y given a certain value $x \in \mathcal{X}$ of the random vector X, one should go along the path which starts from the root and ends in some leaf turning at each parent vertex P to that child P' for which $A_{P'}$ contains x. At the end of the x-specific path, one gets either 1 or −1 which serves as a prediction of Y. Classification tree could be constructed via CART algorithm, see [18].

RF is a non-parametric method of estimating conditional probability $p = p(x)$. Its idea is to improve prediction power of CART tree by taking the average of these trees grown on many bootstrap samples, see [18, ch. 15]. The advantages of this method are low computational costs and the ability to extract relevant predictors when the number of irrelevant ones is large, see [25].

SGB is another non-parametric method of estimating conditional probability $p(x)$. SGB algorithm proceeds iteratively in such a way that, on each step, it builds a new estimate of $p(x)$ and a new classifier decreasing the number of misclassified cases from the previous step, see, e.g., [12].

Standard RF and SGB work poorly for unbalanced samples. One needs either to balance given datasets (as in [26]) before these methods are applied or use special modifications of RF and SGB. To avoid overfitting, permutation test is performed. A common problem of all machine learning methods is a complicated functional form of the final probability estimate $(\text{w.r.t. } x)$ $\hat{p}(x,\xi)$. In genetic studies, one wants to pick up all relevant combinations of SNPs and risk factors, based on a biological pathway causing the disease. Therefore, the final estimate $\hat{p}(x,\xi)$ is to be analyzed. We describe one of possible methods for such analysis within RF framework called conditional variable importance measure (CVIM). One could determine CVIM for each predictor X_i in X and range all X_i in terms of this measure. Following [27], CVIM of predictor X_i given certain subvector Z_i of X is calculated as follows (supposing Z_i takes values $z_{i1},\cdots,z_{im(i)} \in \mathcal{X}^d$ for some $d \in \{1,\cdots,n\}$).

1) For each $k = 1, \cdots, m(i)$, permute randomly the elements of $A_{ik} = \{ j : Z_i^j = z_{ik} \}$ to obtain a vector $\tau^{(k)} := \left(\tau_1^{(k)}, \cdots, \tau_{M(k)}^{(k)} \right)$, where $M(k) = \# A_{ik}$. Consider a vector

$$(l_1, \cdots, l_N) := \left(\tau^{(1)}, \cdots, \tau^{(m(i))} \right).$$

2) Let $B \in \mathbf{N}$. Generate bootstrap samples

$$b = 1, \cdots, B.\ \xi_b = \left(\left(X^{jb}, Y^{jb} \right), j = 1, \cdots, N \right),$$

For each of these samples, construct a CART classifier $f_b(x, \xi_b)$ and calculate

$$CVIM_b = \frac{1}{|C_b|} \sum_{j \in C_b} \left(I\left\{ Y^j = f_b\left(X^j, \xi_b \right) \right\} - I\left\{ Y^j = f_b\left(X^{l_j}, \xi_b \right) \right\} \right)$$

where $C_b = \left\{ j \in \{1, \cdots, N\} : \left(X^j, Y^j \right) \notin \xi_b \right\}$.

3) Compute the final CVIM using the formula

$$CVIM = \frac{1}{B} \sum_{b=1}^{B} CVIM_b \quad .(29)$$

Any permutation $(l_1, \cdots, l_N)$ in the CVIM algorithm destroys dependence between X_i and (Y, Z_{-i}) where Z_{-i} consists of all components of X which are not in Z_i. At the same time it preserves initial empirical distribution of (X_i, Z_i) calculated for the sample x. The average loss of correctly classified Y is calculated, and if it is relatively large w.r.t. CVIM of other predictors, then X_i plays important role in classification and vice versa.

For instance, as Z_i, ($i = 1, \cdots, n$) one can take all the components X_k ($k \neq i$) such that the hypothesis of the independence between X_k and X_i is not rejected at some significance level (e.g., 5%). CVIM-like algorithm could be used to range combinations of predictors w.r.t. the level of association to the disease. This will be published elsewhere.

3. APPLICATIONS: RISKS OF CHD AND MI

We employ here various statistical methods described above to analyze the influence of genetic and non-genetic (conventional) factors on risks of coronary heart disease and myocardial infarction using the data for 454 individuals (333 cases, 121 controls) and 333 individuals (165 cases, 168 controls) respectively. These data contain values of seven SNPs and four conventional risk factors. Namely, we consider glycoprotein Ia (GPIa), connexin- 37 (Cx37), plasminogen activator inhibitor type 1 (PAI-1), glycoprotein IIIa (GPIIIa), blood coagulation factor VII (FVII), coagulation factor XIII (FXIII) and interleukin-6 (IL-6) genes, as well as obesity (Ob), arterial hypertension (AH), smoking (Sm) and hypercholesterolemia (HC). The choice of these SNPs was based on biological considerations. For instance, to illustrate this choice we recall that connexin-37 (Cx37) is a protein that forms gapjunction channels between cells. The mechanism of the SNP in Cx37 gene influence on atherosclerosis development is not fully understood, but some clinical data suggest its importance for CHD development [28]. In the Russian population homozygous genotype can induce MI development, especially in individuals without CHD anamnesis [29].

The age of all individuals in case and control groups ranges from 35 to 55 years to reduce its influence on the risk analysis. For each of considered methods, we use K-fold cross-validation with K = 6. As shown in [16], the standard choice of partition number of cross-validation from 6 to 10 does not change the EPE significantly. We take K = 6 as the sample sizes do not exceed 500. The MSU supercomputer "Chebyshev" was involved to perform computations. As shown below, all applied methods permit to choose appropriate models having EPE for CHD dataset less than 0.25. Thus predictions constructed have significant predictive power. Note that, e.g., in [30] the interplay between genotype and MI development was also studied, with estimated prediction errors 0.30 - 0.40.

3.1. MDR and MDRIR Method

Coronary heart disease. **Table 1** contains EPEs of the most significant combinations of predictors obtained by MDR analysis of coronary heart disease data. Note that we write the gene meaning the corresponding SNP. To estimate the empirical c.d.f. of the prediction error when the disease is not linked with explanatory variables, we used the permutation test. Namely, 100 random uniform permutations of variables $Y^1,\cdots,Y^N$ were generated. For any permutation $(Y_b^1,\cdots,Y_b^N)$ we constructed a sample

$$\bar{\xi}_b = \left(\left(X^1, Y_b^1\right), \cdots, \left(X^N, Y_b^N\right)\right)$$

and applied the same analysis to this simulated sample, here $b = 1, \cdots, 100.$ In these 100 simulations the corresponding empirical prediction error was not less than 0.42. Thus the Monte Carlo p-value of all three combinations was less than 0.01 (since their EPEs were much less than 0.42), which is usually considered as a good performance.

Table 1. The most significant combinations obtained by MDR and MDRIR analysis for CHD and MI data.

Disease	Method	Factors	EPE
CHD	MDR	GPIa, FXIII, AH, HC	0.231
		Cx37, AH, HC	0.238
		GPIa,Cx37, AH, HC	0.241
	MDRIR	FXIII, FVII, AH, HC	0.240
		FXIII, AH, HC	0.242
		GPIa, Cx37, AH, HC	0.247
MI	MDR	GPIIIa, FXIII, Cx37, AH	0.343
		GPIIIa, FXIII, FVII, Cx37	0.347
		Cx37, Sm	0.356
	MDRIR	Cx37, Sm	0.351
		GPIIIa, Cx37, Sm	0.353
		GPIIIa, Cx37, Sm, HC	0.355

Table 1 contains also the results of MDRIR method, which are similar to results of MDR method. However, MDRIR method allows to identify additional combinations (listed in this table) with EPE around 0.24. It follows from the same table that hypertension and hypercholesterolemia are the most important non-genetic risk factors. Indeed, these two factors appear in each of 6 combinations.

To perform a more precise analysis of influence of SNPs on CHD provoking we analyzed gene-gene interactions. We used two different strategies. Namely, we applied MDR method to a subgroup of individuals who were not subject to any of the non-genetic risk factors, i.e. to non-smokers without obesity and without hypercholesterolemia, 51 cases and 97 controls (risk-free sample). Another strategy was to apply MDR method to the whole sample, but to take into account only genetic factors rather than all factors. **Table 2** contains the most significant combinations of SNPs and their EPEs.

Thus based on coronary heart disease data with the help of Tables 1 and 2 we can make the following conclusions. Combination of two SNPs (in GPIa and Cx37) and two non-genetic factors (hypertension and hypercholesterolemia) has the biggest influence on CHD. Also FXIII gives additional predictive power if AH and HC are taken into account.

It turns out that both methods yield similar results. Combination of SNPs in GPIa and Cx37 has the biggest influence on CHD. EPE is about 0.28 - 0.34, and smaller error corresponds to a risk-free sample. Moreover, it follows from Tables 1 and 2 that EPE dropped significantly after additional non-genetic factors were taken into account (the error is 0.247 if additional non-genetic factors are taken into account and 0.343 if not).

Myocardial infarction. EPEs of the most significant combinations obtained by MDR analysis of MI data are presented in **Table 1**. For all 100 simulations of $\overline{\xi_b}$ when the disease was not linked with risk factors, EPE was larger than 0.38. Monte Carlo p-value of all combinations was less than 0.01. MDRIR analysis of the same dataset gave a clearer picture (see the same table), as the pair (Cx37, Sm) appears in all three combinations with the least estimated prediction error.

Apparently, combination of smoking and SNP in Cx37 is the most significant. These two factors appear in all combinations in **Table 1** concerning MI and MDRIR. Involving any additional factors only increases EPE.

Table 2. Comparison of the most significant SNP combinations obtained by two different ways of MDR analysis of CHD data.

Method	Factors	EPE
MDR on individuals not subject to risk factors	GPIa, Cx37	0.281
MDR on the whole group concerning only genetic data	GPIa, Cx37	0.343

The explicit form of the prediction algorithm based on Cx37 and Sm shows that these factors exhibit interaction. Smoking as well as homozygote for recessive allele of the SNP in Cx37 provokes the disease. However wildtype allele can protect from consequences of smoking. Namely, the combination of smoking and Cx37 wildtype is protective, i.e. the value of prediction algorithm of this combination is –1.

3.2. Ternary Logic Regression

We performed several research procedures for CHD and MI data, with different restrictions imposed on the statistical model. Set

$$(X_1,\cdots,X_n)=(Z_1,\cdots,Z_m,R_1,\cdots,R_k)$$

where a vector $Z=(Z_1,\cdots,Z_m)$ stands for SNP values (in PAI-1, GPIa, GPIIIa, FXIII, FVII, IL-6, Cx37 respectively) and $R=(R_1,\cdots,R_k)$ denotes non-genetic risk factors (Ob, AH, Sm, HC), $k=4.\ m=7,$

We considered four different models in order to analyze both total influence of genetic and non-genetic factors and losses in predictive force appearing when some factors were excluded. In our applications we took $s=3$ as search over larger forests for samples with modest sizes could have given very complicated and unreliable results.

Model 1. Define the class M (see Section 2.3) consisting of the functions h having a form

$$h(Z,R)=\beta_0+\sum_{v=1}^{s}\beta_v T_v(Z_1,\cdots,Z_m)+\sum_{v=1}^{k}\beta_{s+v}R_v$$

where $\beta_v \in \mathbf{R}$ and T_v are polynomials identified with trees. In other words we require that non-genetic factors are present only in trees consisting of one variable.

Model 2. Now we assume that any function $h \in M$ has the representation

$$h(Z,R) = \beta_0 + \sum_{v=1}^{s} \beta_v T_v(Z_1,\cdots,Z_m,R_1,\cdots,R_k)$$

where $\beta_v \in \mathbf{R}$ and T_v are polynomials identified with trees. Thus we allow the interaction of genes and nongenetic factors in order to find significant gene-environment interactions. However we impose additional restrictions to avoid too complex combinations of nongenetic risk factors. We do not tackle here effects of interactions where several non-genetic factors are involved. Namely, we consider only the trees satisfying the following two conditions.

1) If there is a leaf containing non-genetic factor variable then the root of that leaf contains product operator.

2) Moreover, another branch growing from the same root is also a leaf and contains a genetic (SNP) variable.

Models 3 and 4 have additional restrictions that polynomials $T_v \ (v=1,\cdots,s)$ in (30) depend only on nongenetic factors and only on SNPs respectively. These models are considered to compare their results with ones obtained with all information taken into account, in order to demonstrate the importance of genetic (resp. non-genetic) data for risk analysis.

Coronary heart disease. The obtained results are provided in **Table 3**.

EPE in Model 1 for CHD was only 0.19. For the same model we performed also fast simulated annealing search of the optimal forest which was much more time-efficient, and a reasonable error of 0.23 was obtained. Model 3 application showed that non-genetic factors play an important role in CHD genesis, as classification based on non-genetic factors only gave the error less than 0.23, while usage of SNPs only (Model 4) let the error grow to 0.34.

Model 1 gave the minimal EPE. For the optimal forest $(T_1,\cdots,R_4)$ the function $\hat{h}(Z,R)$ given before formula (28) with $S=\{1,\cdots,N\}$ is provided by

$$\begin{aligned}&-0.597T_1 - 0.354T_2 + 0.521T_3 - 0.444R_1 \\ &+1.311R_2 - 0.146R_3 + 2.331R_4 - 0.226\end{aligned} \qquad (31)$$

where

$$T_1 = (Z_3 Z_4 + Z_6 Z_7 + (Z_2)^2 + Z_3 Z_7)(Z_1)^2 Z_3 Z_7,$$
$$T_2 = Z_1 (Z_3)^2 (Z_6 Z_7 + Z_2 (Z_4)^2 Z_7),$$
$$T_3 = Z_2 + 2Z_2 (Z_6)^2 Z_7$$

with sums and products modulo 3.

The non-genetic factors 2 and 4 (i.e. AH and HC) are the most influential since the coefficients at them are the greatest ones (1.311 and 2.331). As is shown above, MDR yielded the same conclusion. If the gene-environment interactions were allowed (Model 2), no considerable increase in predictive power has been detected. However we list the pairs of SNPs and non-genetic factors present in the best forest: Z_7 and R_2, Z_7 and R_1, Z_7 and R_4, Z_5 and R_1. We see that SNP in Cx37 is of substantial importance as it appears in combination with all risk factors except for smoking.

As formula (31) is hard to interpret, we select the most significant SNPs via a variant of permutation test. Consider a random rearrangement of the column with first SNP in CHD dataset. Calculate the EPE using these new simulated data and the same function $\hat{h}$ as before. The analogous procedure is done for other columns (containing the values of other SNPs) and the errors found are given in **Table 4** (recall that the EPE equals 0.19 if no permutation is done).

It is seen that the error increases considerably when the values of GPIa and Cx37 are permuted. The statement that they are the main sources of risk agrees with what was obtained above by MDR method.

Myocardial infarction. For the MI dataset, under the same notation that above, the results obtained for our four models are given in **Table 3**. To comment them we should first note that non-genetic risk factors play slightly less important role compared with CHD risk: if they are used without genetic information, the error increases by 0.09, see Models 1 and 3 (while the same increase for CHD was 0.03). The function $\hat{h}(Z,R)$ defined before (28) with $S = \{1, \cdots, N\}$ equals

$$-1.144T_1 + 0.914T_2 - 0.45T_3 - 0.285R_1$$
$$-0.675R_2 + 0.828R_3 - 0.350R_4 - 0.055$$

where $T_1 = Z_1 Z_3 (Z_5)^2$, $T_2 = Z_7$, $T_3 = Z_3 + Z_4 + Z_6 + Z_7$.

Thus the first tree has the greatest weight (coefficient equals –1.144), the second tree (i.e. SNP in Cx37) is on the second place, and non-genetic factors are less important.

Table 3. Results of TLR.

Model	1	2	3	4
EPE for CHD	0.190	0.204	0.228	0.340
EPE for MI	0.305	0.331	0.391	0.365

Table 4. The SNP significance test for CHD in Model 1.

SNP permuted	EPE
GPIa	0.263
Cx37	0.260
IL-6	0.226
PAI-1	0.212
GPIIIa	0.208
FXIII	0.202
FVII	0.190

As for CHD we performed a permutation test to compare the significance of different SNPs. Its results are presented in **Table 5**.

As seen from this table, the elimination of Cx37 SNP leads to a noticeable increase in the EPE. This fact agrees with results obtained by MDR analysis of the same dataset.

Table 5. The SNP significance test for MI in Model 1.

SNP permuted	EPE
Cx37	0.444
GPIIIa	0.353
IL-6	0.340
FXIII	0.328
FVII	0.324
PAI-1	0.319
GPIa	0.305

3.3. Results Obtained by RF and SGB Methods

Since machine learning methods give too complicated estimates of the dependence structure between Y and X, we have two natural ways to compare them with our methods. Namely, these are the prediction error and the final significance of each predictor. The given datasets were unbalanced w.r.t. response variable and we first applied the resampling technique to them. That means enlargement of the smaller of two groups case-control in the sample by additional bootstrap observations until the final proportion case:control is 1:1. Note that due to the resampling techniques the following effect arises: some observations in small groups (case or control) appear in the new sample more frequently than other ones. Therefore, we took the average over 1000 iterations.

Coronary heart disease. Results of RF and SGB methods are given in **Table 6**. It shows that RF and SGB methods gave statistically reliable results (EPE in the permutation test is close to 50%). Moreover, additional SNP information improved predicting ability by 13% (SGB). It seems that SGB method is fitted better to CHD data than RF.

To compute CVIM for each X_i, we constructed a vector Z_i as follows. Let Z_i contain all predictors X_j, $j \neq i$, for which chi-square criteria rejected independence hypothesis between X_j and X_i at 5% significance level. **Table 7** shows that the most relevant predictors for CHD are AH, HC and Cx37.

Table 6. EPE/EPE in permutation test calculated via crossvalidation for CHD and MI datasets with employment of RF and SGB methods.

Disease	Data	RF	SGB
CHD	with SNP	0.200/0.454	0.134/0.473
	without SNP	0.230/0.510	0.261/0.503
MI	with SNP	0.360/0.497	0.399/0.530
	without SNP	0.473/0.527	0.482/0.562

Table 7. Predictors ranged in terms of their CVIM for CHD and MI dataset.

CHD		MI	
AH	8.90	Cx37	7.50
HC	5.30	Sm	2.00
Cx37	5.10	AH	1.86
Ob	0.56	GPIIIa	0.03
FXIII	0.53	FVII	0.02
Sm	0.11	FXIII	≈0
GPIa	0.10	HC	≈0
FVII	0.07	GPIa	≈0
PAI-1	0.03	Ob	≈0
GPIIIa	0.02	IL-6	≈0
IL-6	0.01	PAI-1	≈0

Myocardial infarction. Results of RF and SGB methods are given in **Table 6**. It shows that RF and SGB methods gave statistically reliable estimates (EPE in the permutation test is close to 50%). Moreover, additional SNP information improved predicting ability by 11% (RF method).

CVIM was calculated according to (29) and is given in **Table 7**. Thus, the most relevant predictors for MI are Cx37, Sm and AH.

4. CONCLUSIONS AND FINAL REMARKS

In the current study we developed important statistical methods concerning the analysis of multidimensional genetic data. Namely, we proposed the MDR with independent rule and ternary logic regression with a new version of simulated annealing. We compared them with several popular methods which appeared during the last decade. It is worth to emphasize that all considered methods yielded similar qualitative results for dataset under study.

Let us briefly summarize the main results obtained. The analysis of CHD dataset showed that two non-genetic risk factors out of four considered (AH and HC) had a strong connection with the disease risk (the error of classification based on non-genetic factors only is 0.25 - 0.26 with p-value less than 0.01). Also, the classification based on SNPs only gave an error of 0.28 which is close to one obtained by means of non-genetic predictors. Moreover, the most influential SNPs were in genes Cx37 and GPIa (FXIII also entered the analysis only when AH and HC were present). EPE decreased to 0.13 when both SNP information and non-genetic risk factors were taken into account and SGB was employed. Note that excludeing any of the 5 remaining SNPs (all except for two most influential) from data increased the error by 0.01 - 0.02 approximately. So, while the most influential data were responsible for the situation within a large part of population, there were smaller parts where other SNPs came to effect and provided a more efficient prognosis ("small subgroups effect"). The significance of SNP in GPIa and FXIII genes was observed in our work. AH and HC influence the disease risk by affecting the vascular wall, while GPIa and FXIII may improve prognosis accuracy because they introduce haemostatic aspect into analysis.

The MI dataset gave the following results. The most significant factors of MI risk were the SNP in Cx37 (more precisely, homozygous for recessive allele) and smoking with a considerable gene-environment interacttion present. The smallest EPE of methods applied was 0.33 - 0.35 (with p-value less than 0.01). The classification based on non-genetic factors only yielded a greater error of 0.42. Thus genetic data improved the prognosis quality noticeably. While two factors were important, other SNPs considered actually did not improve the prognosis essentially, i.e. no small groups effect was observed.

While CHD data used in the study permitted to specify the most important predictors with EPE about 0.13, the MI data lead to less exact prognoses. Perhaps this complex disease requires a more detailed description of individual's genetic characteristics and environmental factors.

The conclusions given above are based on several complementary methods of modern statistical analysis. These new data mining methods allow to analyze other datasets as well. The study can be continued and the medical conclusions need to be replicated with larger datasets, in particular, involving new SNP data.

5. ACKNOWLEDGEMENTS

The work is partially supported by RFBR grant (project 10-01-00397a).

REFERENCES

1. Y. Fujikoshi, R. Shimizu and V. V. Ulyanov, "Multivariate Statistics: High-Dimensional and Large-Sample Approximations," Wiley, Hoboken, 2010.
2. S. Szymczak, J. Biernacka, H. Cordell, O. GonzálezRecio, I. König, H. Zhang and Y. Sun, "Machine Learning in Genome-Wide Association Studies," Genetic Epidemiology, Vol. 33, No. S1, 2009, pp. 51-57.
3. D. Brinza, M. Schultz, G. Tesler and V. Bafna, "RAPID Detection of Gene-Gene Interaction in Genome-Wide Association Studies", Bioinformatics, Vol. 26, No. 22, 2010, pp. 2856-2862.
4. K. Wang, S. P. Dickson, C. A. Stolle, I. D. Krantz, D. B. Goldstein and H. Hakonarson, "Interpretation of Association Signals and Identification of Causal Variants from Genome-Wide Association Studies," The American Journal of Human Genetics, Vol. 86, No. 5, 2010, pp. 730-742.
5. Y. Liang and A. Kelemen. "Statistical Advances and Challenges for Analyzing Correlated High Dimensional SNP Data in Genomic Study for Complex Diseases," Statistics Surveys, Vol. 2, No. 1, 2008, pp. 43-60.
6. H. Schwender and I. Ruczinski, "Testing SNPs and Sets of SNPs for Importance in Association Studies," Biostatistics, Vol. 12, No. 1, 2011, pp. 18-32.
7. M. Ritchie, L. Hahn, N. Roodi, R. Bailey, W. Dupont, F. Parl and J. Moore, "Multifactor-Dimensionality Reduction Reveals High-Order Interactions among EstrogenMetabolism Genes in Sporadic Breast

Cancer," The American Journal of Human Genetics, Vol. 69, No. 1, 2001, pp. 138-147.

8. D. Velez, B. White, A. Motsinger, W. Bush, M. Ritchie, S. Williams and J. Moore, "A Balanced Accuracy Function for Epistasis Modeling in Imbalanced Datasets Using Multifactor Dimensionality Reduction," Genetic Epidemiology, Vol. 31, No. 4, 2007, pp. 306-315.

9. I. Ruczinski, C. Kooperberg and M. LeBlanc, "Logic Regression," Journal of Computational and Graphical Statiststics, Vol. 12, No. 3, 2003, pp. 475-511.

10. H. Schwender and K. Ickstadt, "Identification of SNP Interactions Using Logic Regression," Biostatistics, Vol. 9, No. 1, 2008, pp. 187-198.

11. L. Breiman, "Random Forests," Machine Learning, Vol. 45, No. 1, 2001, pp. 5-32.

12. J. Friedman, "Stochastic Gradient Boosting," Computational Statistics & Data analysis, Vol. 38, No. 4, 2002, pp. 367-378.

13. X. Wan, C. Yang, Q. Yang, H. Xue, N. Tang and W. Yu, "Mega SNP Hunter: A Learning Approach to Detect Disease Predisposition SNPs and High Level Interactions in Genome Wide Association Study," BMC Bioinformatics, Vol. 10, 2009, p. 13.

14. A. Bulinski, O. Butkovsky, A. Shashkin, P. Yaskov, M. Atroshchenko and A. Khaplanov, "Statistical Methods of SNPs Analysis," Technical Report, 2010, pp. 1-159 (in Russian).

15. G. Bradley-Smith, S. Hope, H. V. Firth and J. A. Hurst, "Oxford Handbook of Genetics," Oxford University Press, New York, 2010.

16. S. Winham, A. Slater and A. Motsinger-Reif, "A Comparison of Internal Validation Techniques for Multifactor Dimensionality Reduction," BMC Bioinformatics, Vol. 11, 2010, p. 394.

17. A. Arlot and A. Celisse, "A Survey of Cross-Validation Procedures for Model Selection," Statistics Surveys, Vol. 4, No. 1, 2010, pp. 40-79.

18. T. Hastie, R. Tibshirani and J. Friedman, "The Elements of Statistical Learning: Data Mining, Inference, and Prediction," 2nd Edition, Springer, New York, 2009.

19. R. L. Taylor and T.-C. Hu, "Strong Laws of Large Numbers for Arrays of Rowwise Independent Random Elements," International Journal of Mathematics and Mathematical Sciences, Vol. 10, No. 4, 1987, pp. 805-814.

20. E. Lehmann and J. Romano, "Testing Statistical Hypotheses," Springer, New York, 2005.

21. P. Golland, F. Liang, S. Mukherjee and D. Panchenko, "Permutation Tests for Classification," Lecture Notes in Computer Science, Vol. 3559, 2005, pp. 501-515.

22. J. Park, "Independent Rule in Classification of Multivariate Binary Data," Journal of Multivariate Analysis, Vol. 100, No. 10, 2009, pp. 2270-2286.

23. S. Lee, Y. Chung, R. Elston, Y. Kim and T. Park, "LogLinear Model-Based Multifactor Dimensionality Reduction Method to Detect Gene-gene Interactions," Bioinformatics, Vol. 23, No. 19, 2007, pp. 2589-2595.

24. A. Nikolaev and S. Jacobson, "Simulated Annealing," In: M. Gendreau and J.-Y. Potvin, Eds., Handbook of Metaheuristics, Springer, New York, 2010, pp. 1-39.

25. G. Biau, "Analysis of a Random Forests Model," LSTA, LPMA, Paris, 2010.

26. N. Chawla, "Data Mining for Imbalanced Datasets: An Overview," In: O. Maimon and L. Rokach, Eds., Data Mining and Knowledge Discovery Handbook, Springer, New York, 2010, pp. 875-886.

27. C. Strobl, A. Boulesteix, T. Kneib, T. Augustin and A. Zeileis, "Conditional Variable Importance for Random Forests," BMC Bioinformatics, Vol. 9, 2008, p. 307.

28. A. Hirashiki, Y. Yamada, Y. Murase, Y. Suzuki, H. Kataoka, Y. Morimoto, T. Tajika, T. Murohara and M. Yokota, "Association of Gene Polymorphisms with Coronary Artery Disease in Lowor High-Risk Subjects Defined by Conventional Risk Factors," Journal of the American College of Cardiology, Vol. 42, No. 8, 2003, pp. 1429-1437.

29. A. Balatskiy, E. Andreenko and L. Samokhodskaya, "The Connexin37 Polymorphism as a New Risk Factor of MI Development," Siberian Medical Journal, Vol. 25, No. 2, 2010, pp. 64-65 (in Russian).

30. C. Coffey, P. Hebert, M. Ritchie, H. Krumholz, J. Gaziano, P. Ridker, N. Brown, D. Vaughan and J. Moore, "An Application of Conditional Logistic Regression and Multifactor Dimensionality Reduction for Detecting GeneGene Interactions on Risk of Myocardial Infarction: The Importance of Model Validation," BMC Bioinformatics, Vol. 5, 2004, p. 49.

Chapter 6

A Comparison of Statistics for Assessing Model Invariance in Latent Class Analysis

Holmes Finch

Department of Educational Psychology, Ball State University, Muncie, USA

ABSTRACT

Latent class analysis (LCA) is a widely used statistical technique for identifying subgroups in the population based upon multiple indicator variables. It has a number of advantages over other unsupervised grouping procedures such as cluster analysis, including stronger theoretical underpinnings, more clearly defined measures of model fit, and the ability to conduct confirmatory analyses. In addition, it is possible to ascertain whether an LCA solution is equally applicable to multiple known groups, using invariance assessment techniques. This study compared the effectiveness of multiple statistics for detecting group LCA invariance, including a chi-square difference test, a bootstrap likelihood ratio test, and several information indices. Results of the simulation study found that the bootstrap likelihood ratio test was the optimal invariance assessment statistic. In addition to the simulation, LCA group invariance assessment was demonstrated in an application with the Youth Risk Behavior Survey (YRBS). Implications of the simulation results for practice are discussed.

Keywords: Latent Class Analysis, Model Invariance, Information Indices

1. INTRODUCTION

Over the last 20 years, latent class analysis (LCA) has become a very popular statistical method for identifying unobserved groups in the population using multiple observed dichotomous or polytomous outcome variables. A variant of this technique, commonly referred to as latent profile analysis (LPA), can be used to identify latent classes with observed continuous outcomes. LCA is used in both exploratory [1] and confirmatory [2] analyses. In the exploratory case, researchers have no testable a priori hypotheses regarding the nature of the latent classes in the population. It is certainly possible that they would have some expectations as to the number of such classes, but these surmises would not be sufficiently strong to warrant a formal statement of their existence in the form of hypotheses [3] . In contrast, with confirmatory LCA the researcher would typically have hypotheses about the number of latent classes, as well as some characteristics of these classes in the form of absolute or relative response probabilities on the observed variables. They would then test these hypotheses by comparing models with different parameter constraints [4] . Another application of LCA allows for the comparison of latent class solutions across known groups (e.g. gender) within the broader population [2] . For example, Collins and Lanza [5] have demonstrated the application of multiple groups LCA (MGLCA) as a means for comparing patterns of delinquency across different adolescent age groups. Their work demonstrates how MGLCA can be used to assess the invariance of latent class solutions across two or more known subgroups. The primary purpose of this simulation study is to extend their work by comparing the performance of several statistics that have been suggested for use in assessing the null hypothesis of latent class invariance across multiple known groups.

1.1. The LCA Model

The basic LCA model is described in some detail by McCutcheon [2] . Assume that data have been collected for four observed, dichotomous variables, X1, X2, X3, and X4, and that there exists a latent categorical variable Y, which accounts for the relationships among these four observed variables. The LCA model linking the latent and observed variables can then be expressed as:

$$\pi_{ijklt}^{X1X2X3X4y} = \pi_t^Y \pi_{it}^{X1|Y} \pi_{jt}^{X2|Y} \pi_{kt}^{X3|Y} \pi_{lt}^{X4|Y} \quad (1)$$

where

$\pi_t^Y =$ Probability that a randomly selected individual will in latent class cof latent variable Y ;

$\pi_{it}^{X1|Y} =$ Probability that a member of latent class t will provide a response of i for variable X1;

$\pi_{it}^{X2|Y} =$ Probability that a member of latent class t will provide a response of i for variable X2;

$\pi_{it}^{X3|Y} =$ Probability that a member of latent class t will provide a response of i for variable X3;

$\pi_{it}^{X4|Y} =$ Probability that a member of latent class t will provide a response of i for variable X4.

As an example of how these parameters are interpreted, take an individual from the population who has the following probability values for the three classes in Y: $\pi_1^Y = 0.6$, $\pi_2^Y = 0.25$, and $\pi_3^Y = 0.15$. These results indicate that the individual is most likely a member of class 1 of the latent variable, with only a 1/4 chance of being in class 2 and a less than 1/5 chance of being in class 3. In addition, assume that observed variable X1 is a survey item asking whether an individual never uses sunscreen when they go outside for extended periods of time. A value for $\pi_{\text{yes\_1}}^{X1|Y}$ of 0.25 would indicate that an individual in the first class of the latent variable would have a fairly low likelihood of responding Yes to this item; i.e. most members of latent class 1 do use sunscreen when spending extended periods of time outside.

1.2. MG LCA and Invariance Testing

As mentioned earlier, in some instances a researcher may be interested in ascertaining whether the nature of latent classes differs across known subgroups in the populations. For example, psychologists studying risk taking behavior among adolescents may have theoretical reasons to believe that in the broad teen aged population there are a relatively small number of latent classes with specific risk taking profiles. Further, it may be hypothesized that these classes differ for males and females, either in terms of their relative frequency, their item response profiles or both. In this context, the researcher would like to determine 1) whether indeed coherent latent classes of youth exist based on risk taking behavior, and if so, 2) whether these classes differ for males and females. It is for this type of problem that MGLCA is appropriate.

Collins and Lanza [5] provide an excellent and detailed discussion of invariance testing using MGLCA. There- fore, this manuscript contains only a summary of the methodology, and the interested reader who would like more detail is referred to [5] . The MGLCA model is simply an extension of the original LCA model in (1):

$$\pi_{ijkltg}^{X1X2X3X4y} = \pi_{tg}^{YG} \pi_{itg}^{X1|YG} \pi_{jtg}^{X2|YG} \pi_{ktg}^{X3|YG} \pi_{ltg}^{X4|YG} \quad (2)$$

where all terms are as defined previously and the new term G $(g = 1, \cdots, G)$ refers to observed group membership (e.g. gender). In other words, latent class prevalence and item response probabilities are conditional on group membership, thus allowing for the possibility of differences across the groups.

The researcher interested in group invariance for the LCA models will want to assess whether the number of latent classes, the class specific item response probabilities and the prevalence of the classes are the same across the known groups of interest. Typically LCA invariance testing begins by determining the number of latent classes in the population as a whole, followed by a determination as to whether the number is the same in each group. This stage of the analysis is very akin to assessing configural invariance in the factor analysis context, where no equality constraints are placed on parameters, and only the groups' factor patterns are compared using exploratory factor analysis. Similarly, in LCA the researcher would determine whether the number of latent classes is the same across groups, using exploratory LCA with each group independently. This is typically done by fitting LCA models to the groups separately and using standard methods for determining the number of latent classes in each. Selecting the optimal model for a given dataset is typically done using one or more goodness of fit statistics [6] .

If it has been determined that the number of latent classes is equivalent across groups, the researcher may then go on to assess whether item response probabilities are also equal across groups. This stage of LCA invariance assessment corresponds loosely to measurement invariance assessment in factor analysis, forming the second step in the process. In factor analysis, measurement invariance involves the determination as to whether factor loadings are equivalent across groups, while measurement invariance in LCA refers to whether performance on the observed indicators (e.g. probability of endorsing an item) is equivalent across groups. However, if the num- ber of classes differs across groups, no further invariance assessment should be conducted [5] . With factor analysis, a lack of measurement invariance (i.e. group differences in factor loadings) means that the factors themselves have a different interpretation across the groups. Similarly, in the context of LCA a lack of measurement invariance indicates that the nature of the latent classes differs across the groups. If the classes have different response probabilities for one or more indicator variables across the known groups, then the meaning of the classes differs across the groups as well, and so must the interpretation of them. In such a case, researchers cannot describe the population as consisting of a set number of latent classes with common traits, but rather must think of the classes as being fundamentally different in nature depending on one's membership in a given known group.

In order to determine whether the latent class specific response probabilities are invariant across groups, the researcher would fit two models to the data. In the first model all response probabilities are allowed to vary across groups, while in the second model these probabilities are constrained to be equal. Using the example described previously, in the second model the response probability for variable X1 for latent class 1 is constrained to be the same for both groups, as are the response probabilities for item X1 in latent class 2, item X2 in latent class 1 and so on. The fit of the two models to the data is then compared using one or more of the statistical techniques described below. If the constrained model provides a substantially worse fit than the unconstrained, the researcher would conclude that the groups differ in terms of one or more of the latent class specific item response probabilities. As an example, the researcher interested in comparing latent classes based on risk behavior across genders would first fit a model in which all of the latent class specific item response probabilities were free to vary between males and females, and then a second model in which they were constrained to be equal for the two groups. She would then compare the fit of the two models to determine whether the LCA model for risk was invariant across genders.

If the number of latent classes and the item response probabilities are both determined to be invariant across groups, the researcher can test for the invariance of class prevalence. In other words, she will assess whether the relative size of each latent class is the same across the observed groups. This third stage in the invariance testing methodology is analogous to testing for metric invariance in factor analysis, in that we are interested in the extent to which the mean group prevalence is the same across groups, in the same way that we are interested in whether indicator intercepts are group equivalent with metric invariance. Comparing the equivalence of class prevalence across groups is done in much the same manner as assessing response probability invariance: A model constraining response probabilities and allowing prevalence rates to vary is fit to the data, and its fit is compared to that of the fully constrained model. If the constrained model provides substantially worse fit than that of the unconstrained prevalence model, then the researcher would conclude that class prevalence rates are not invariant across groups.

1.3. Statistics Used in Assessing LCA Model Invariance

As described above, the process of assessing LCA model invariance involves the comparison of fit for a fully constrained and several unconstrained models. This comparison of relative fit is key to the success of the entire invariance testing venture, and thus is of some importance to researchers interested in this type of analysis. A number of statistical tools have been suggested for use in the assessment of LCA invariance, including the likelihood ratio test, as well as a variety of information indices. However, there has been little research focused on the relative performance of these statistics in terms of making accurate determinations regarding the presence (or absence) of LCA model invariance. Therefore, as mentioned previously, the primary focus of this study is on the systematic comparison of these statistical techniques in order to ascertain which might be most accurate in the context of LCA invariance assessment, both in terms of correctly identifying when invariance is not present, as well as when it is.

The statistics to be examined in this study can be divided into three broad classes: a chi-square difference test, information indices, and a parametric bootstrap test. The first of these is based on the likelihood ratio statistic, (G^2), which is a measure of absolute fit in the single group LCA context, and can be used to construct a test of invariance in the multiple groups situation. With a single group, G^2 tests the null hypothesis that the proposed latent class solution adequately fits the observed data, and can be expressed as

$$G^2 = 2\sum_{w=1}^{W} f_w \ln\left(\frac{f_w}{\hat{f}_w}\right) \tag{3}$$

where

f_w = Observed frequency in cell w

$\hat{f}_w$ = Model predicted frequency in cell w

The G^2 statistic essentially compares the actual counts in each of the W cells created by the set of categorical indicator variables with the cell counts predicted by the model. The further the predicted counts are from the actual, the greater value of the test statistic and the more evidence exists that the proposed model does not fit the observed data. The degrees of freedom for this test are W-P-1, where W is the total number of cells obtained from the cross- tabulation of the observed variables, and P is the number of parameters estimated by the model. G^2 can be compared to the χ^2 distribution with W-P-1 degrees of freedom under certain conditions. This test statistic relies on the assumption of multivariate normality of the observed indicator variables, which is not often met in practice, in which case the G^2 may not be a reliable indicator of model fit [2] [6] .

In the context of multiple groups LCA, which is the focus of this study, the difference of G^2 statistics can be used to assess the null hypothesis of LCA parameter invariance. For example, consider the case in which we are interested in assessing the measurement invariance of a LCA solution across two groups. Two models would need to be fit to the data, the first allowing item parameter values to vary across the two groups, and the second constraining these parameters to be equal across the groups. The relative fit of these two models can then be compared using the difference between their G^2 values. This G^2_Δ statistic is asymptotically distributed as a χ^2 with $df_\Delta = df_{\text{model2}} - df_{\text{model1}}$. A statistically significant result for this test would indicate that the fit of the two models differed; i.e. constraining the model parameters to be equal between the groups resulted in worse model fit than allowing them to vary [5] . This strategy is very similar to that for testing the fit of two structural equation models. Indeed, the use of the G^2_Δ difference test for invariance assessment in the context of structural equation modeling (SEM) has been studied by a number of researchers and found to be an effective tool for accurately identifying the presence of noninvariance in many situations [7] -[9] .

An alternative approach to assessing invariance using the G^2_Δ test involves a class of statistics known collectively as information criteria (IC). These IC's are all based on the log-likelihood of a given model, each applying a different penalty for model complexity. The smaller the value of an IC, the better fitting the model is said to be. In the context of LCA, several such criteria are used to assess model fit, including the Akaike Information Criterion (AIC), the consistent AIC (CAIC), the Bayesian Information Criterion (BIC), and the sample size adjusted BIC (aBIC). The AIC is calculated as:

$$\mathrm{AIC} = -2\ln(L) + 2p \tag{4}$$

where

$\ln(L)$ is the log likelihood for the model, and p is the number of estimated model parameters [10] . Thus, holding the value of $\ln(L)$ constant, models with more parameters will have larger values of AIC. An alternative to the AIC, which includes both the number of model parameters and the sample size (N) in the penalty term is the CAIC [11] .

$$\mathrm{CAIC} = -2\ln(L) + p*(1+\ln(N)) \tag{5}$$

The BIC takes a similar form to the CAIC, and is calculated as:

$$\mathrm{BIC} = -2\ln(L) + p*\ln(N) \tag{6}$$

Finally, the aBIC (Sclove, 1987) involves an adjusted sample size value.

$$\mathrm{aBIC} = -2\ln(L) + p*\ln(n^*) \tag{7}$$

Where

$$n^* = (N+2)/24$$

A third general approach for assessing the relative fit of the constrained and unconstrained models that will be investigated here is based upon the parametric bootstrap likelihood ratio test that has been shown effective for identifying the correct number of classes in LCA [12] . When applied to the problem of identifying the correct number of classes for standard one group LCA, this methodology involves the estimation of the distribution of the difference in G^2 values when comparing models that differ in terms of the number of latent classes. In other words, a researcher interested in comparing the fit of k and $k-1$ latent classes would use the BLRT to obtain a p-value that could be used to determine whether the fit was significantly different between models for the two numbers of latent classes. A primary advantage of this approach is that it does not rely on assumptions regarding the distribution of the difference in G^2 values for the k and $k-1$ models [12] .

In the current application, the BLRT was used to assess the difference in fit for the constrained and unconstrained models described above. The basic methodology is very similar to that outlined in [6] , in the following steps:

1) Estimate the constrained and unconstrained models, and obtain the G^2 values for each.

2) Under the null model that invariance holds (i.e. the constrained model), generate a bootstrap sample and calculate the difference in G^2 values $\left(G^2_\Delta\right)$ for the constrained and unconstrained models.

3) Repeat step 2 B (e.g. 1000) times in order to obtain the distribution of G^2_Δ when the null hypothesis of invariance holds.

4) Obtain the p-value testing the invariance hypothesis by comparing the actual G^2_Δ statistic with the distribution from step 3. If the observed value is greater than or equal to the 95th percentile of the distribution, then reject the null hypothesis of model invariance. The BLRT can be applied to the testing of both measurement invariance, and the equivalence of class prevalences across groups.

1.4. Prior Research in LCA Model Selection and Model Invariance Assessment

While, as mentioned previously, there has been relatively little work in assessing the performance of statistics for testing model invariance in the context of LCA, research has been conducted investigating methods for invariance assessment of structural equation models, and comparing approaches for identifying optimal LCA models. These studies provide insights into the broader questions of invariance assessment and LCA model selection, and will therefore be reviewed here.

With regard to the selection of the optimal number of classes in various types of mixture models, researchers have compared a number of methods, including information indices such as the AIC, BIC, and aBIC, along with hypothesis tests including the G^2_Δ test, and the BLRT. These earlier studies generally found that among the information indices, the aBIC was optimal for correctly selecting the number of latent classes for growth mixture models [13] , and for latent class models [14] . Other research has found that the AIC can lead to overestimation of the number of latent classes [15] , while the BIC provides reasonably accurate determination as to the number of classes [16] . In terms of hypothesis tests, research has revealed that for small samples or a large number of indicator variables, the G^2_Δ test tends to overestimate the number of classes, because the underlying distribution of G^2_Δ difference values does not follow the chi-square [17] -[19] . Further work in this area [6] replicated this overestimation problem for the G^2_Δ test. In addition, their results showed that the BLRT was able to maintain the nominal Type I error rate, while yielding optimal power for identifying the correct number of latent classes. Nylund et al. [6] also found that the aBIC was consistently the most accurate of the information indices in terms of identifying the correct number of latent classes when the indicator variables were categorical, as is the case in the current study, and that the CAIC was less likely than the AIC to overestimate the number of classes. Finally, when considering all statistical methods for determining the number of latent classes, Nylund et al. concluded that the BLRT was the optimal method for LCA model selection, overall, even when compared to the aBIC. Thus, one goal of the current study was to ascertain whether this approach would also be effective for the problem of model invariance assessment.

The literature on invariance testing for structural equation modeling (SEM) has also focused on the use of both hypothesis testing and information indices. For example, French and Finch [7] found that when indicator variables were normally distributed continuous variables, the G^2_Δ test exhibited good control of the Type I error rate when testing for measurement invariance (group differences in loadings), while at the same time yielding relatively high power for identifying a lack of invariance. However, when the indicators were dichotomous, they found that this test generally had very low power across sample sizes. Meade, and Lautenschlager [20] reported that with large samples (2000 or more), the G^2_Δ test had high power for detecting even minor differences in the constrained and unconstrained models, leading to findings of noninvariance for very small actual differences in parameters across groups. With respect to information indices and SEM invariance assessment, Burnham and Anderson [21] suggested use of the CAIC over the AIC when the ratio of sample size to number of parameters was less than 40, similar to the results reported in Nylund et al. in the context of LCA. On the other hand, Vrieze [22] concluded that for complex latent variable models, the AIC may be the preferable approach because of its particular min-max properties not shared by other information indices, such as the BIC. In a similar vein, Shao [23] suggested that AIC is preferable to BIC when the true model in the population is very complex, which he argued is probably most often the case in practice. Focusing specifically on factor invariance assessment with information indices, Wicherts and Dolan [24] reported that these indices can be very useful for selecting between constrained and unconstrained models, as long as researchers are careful to ensure that the two models are identical with the exception of the parameters being tested. They noted that when this is not the case, invariance assessment will be compromised. In short, while there has been much work investigating approaches for model invariance assessment in SEM, the results have not produced a clearly optimal approach, unlike the BLRT for LCA model selection. The G^2_Δ test was not effective for categorical indicator variables, which are common in LCA, and no single information index consistently performed the best for SEM invariance assessment. Thus, it is hoped that the current work may expand this literature by both investigating the performance of these traditional methods of invariance assessment in the context of LCA, rather than SEM, and by comparing them with BLRT, which has heretofore been used for LCA model selection rather than invariance testing.

1.5. Goal of the Current Study

The primary goal of the current study was to compare the performance of a variety of statistical approaches for assessing group invariance in the context of LCA. While much prior work has been done comparing methods for assessing overall model fit in the context of LCA, there is very little published research comparing statistics for assessing latent class model invariance across multiple known groups. In addition, while these statistics are all familiar parts of the statistics landscape and have been studied in other modeling contexts, they have not been examined in terms of their abilities to accurately assess latent class invariance. Therefore, these methods, including the G^2_Δ test, BLRT, AIC, BIC, CAIC and aBIC, were used to check for the presence of measurement invariance and to compare class prevalences across two known groups, under a variety of simulated conditions. Based on the prior literature that is described above, it is hypothesized that the BLRT will generally perform well, given its aforementioned strength in correctly identifying the number of latent classes in the population. It is also anticipated that, given prior work demonstrating its sensitivity to both non-normality and sample size, the G^2_Δ test will have difficulty with the invariance problem studied here, given the nonnormal (dichotomous) nature of the indicator variables, and the variety of sample sizes studied. Finally, as Collins and Lanza [5] have noted, the information indices may be better suited to narrowing down the list of possible optimal models rather than identifying a single best model. Among these statistics, prior studies provide some evidence that the aBIC may be a top choice, but much of the work supporting this hypothesis has been limited to the continuous indicator case. In sum, as has been noted in earlier writing [21] [22] , more work needs to be done investigating the relative performance of a number of model selection tools in the context of mixture models, including LCA invariance assessment. In addition to using a simulation approach to compare the performance of invariance testing statistics under a variety of conditions, a secondary goal of this study was to demonstrate the practice of invariance assessment using MGLCA with data taken from the Youth Risk Behavior Survey (YRBS). The results of both the simulation and applied data analysis should serve to inform both practitioners and researchers regarding the use of MGLCA for invariance assessment, and should further work in the area of LCA invariance assessment.

2. METHODS

Data for this Monte Carlo study were generated using the Simulate Lca Dataset SAS macro, version 1.1.0 [25] , while data were analyzed using PROC LCA [26] under the SAS version 9.1 system [27] . All manipulated factors, which are described in detail below, were crossed such that there were 6 (latent class assessment procedure) × 2 (type of invariance) × 2 (number of latent classes) × 4 (sample size) × 2 (observed group size ratio) × 2 (latent class size ratio) × 3 (proportion of noninvariant items) × 4 (degree of noninvariance across groups) combinations, for a total of 4608 individual study conditions. For each combination of conditions, 1000 simulation replications were generated and analyzed. As mentioned previously, the approaches to invariance assessment included in this study were the G^2_Δ test, BLRT, AIC, BIC, CAIC and aBIC. The manipulated simulation conditions, each of which is described in detail below, were based upon data reported in Collins and Lanza [5] , chapter 5. For all of the conditions, 2 groups were simulated with 5 dichotomous indicator variables. The data were simulated using the pattern of item endorsement probabilities that appear in Table 1. Note that these values were identical for the two observed groups in the invariant condition, while they changed for group 2 in the noninvariant cases, as described below. For the 2 latent class conditions only the item endorsement probabilities in the first two columns of Table 1 were used.

2.1. Manipulated Study Conditions

Type of Invariance

Two types of invariance were assessed in this study. Measurement invariance involved testing whether the latent class specific item response probabilities were the same between the two known groups. In addition, thee quivalence of the relative frequencies of the latent classes in each group was also tested. Based upon recommendations in Collins and Lanza [5] , the equality of the latent class prevalences across groups was tested under the assumption of strict measurement invariance. Thus, item response probabilities for the latent classes were simulated to be the same across groups when latent class prevalences were compared.

2.2. Number of Latent Classes

Data were simulated with either 2 or 4 latent classes in each observed group for the measurement invariance analysis. For testing the equivalence of class prevalence, 2 latent classes were simulated for each observed group, in order to allow for focus on the basic test of invariance without the complicating presence of multiple groups. For all simulation conditions, each group contained the same number of latent classes. It is recognized that in actual practice the number of latent classes will not always be the same for multiple groups in such an invariance study, making this design decision a limitation of the current study. However, this decision was made for two reasons. First, this simulation setup reflects the case where the first level of invariance has been met, namely that the number of groups in the latent classes is identical. Were this condition not to hold, further invariance testing would not be conducted, as the researcher would presumably have discovered that the number of classes differed across groups when he fit the LCA models separately for them. The second reason for this design condition is to keep the size of the study and number of conditions manageable. Nonetheless, it is certainly true that keeping the number of classes constant for the two groups does not reflect every condition that a researcher might see in practice, and is therefore a limitation of the current study.

Table 1. Endorsement probabilities used to simulate data, by item and latent class.

Item	Latent class 1	Latent class 2	Latent class 3	Latent class 4
1	0.8	0.2	0.2	0.8
2	0.7	0.3	0.2	0.8
3	0.6	0.4	0.3	0.7
4	0.3	0.7	0.3	0.7
5	0.2	0.8	0.1	0.9

2.3. Sample Size and Observed Group Size Ratio

Four total sample size conditions were simulated, including 200, 500, 1000 and 2000, which were designed to represent a range of values that appear in the applied literature in which LCA is used. Two observed group size ratio conditions were also simulated: 1/1 and 3/1.

2.4. Latent Class Size Ratio

For the measurement invariance aspect of the study, the ratio of latent classes within each group were simulated to be either 50/50 or 70/30 in the 2 latent classes condition. In the 4 class case, the ratios were either 25/25/25/25 or 55/15/15/15. In the results section below, we will refer to the equal ratio condition as 50/50 for both 2 and 4 latent classes, and the unequal ratio condition as 70/30 for both numbers of latent classes. We recognize that there are many other possible latent class ratios that could be included. However, in order to keep the size of the study manageable we have elected to examine only these conditions. For the portion of the study examining latent class prevalence, the latent class ratios were manipulated as described in the Degree of noninvariance section below.

2.5. Proportion of Noninvariant Items

For the measurement invariance aspect of the study, 3 conditions for proportion of noninvariant items were simulated, including 0% (complete invariance), 20% (1 target item) and 40% (2 target items).

2.6. Magnitude of Measurement Noninvariance

The magnitude of noninvariance present in the target items was based loosely on results regarding adolescent delinquency reported in Collins and Lanza [5] . The magnitude of noninvariance between the observed groupslatent class specific item response probabilities for target items were set to 0 (complete invariance corresponding to the 0% proportion of invariant item condition described above), 0.1, 0.3 or 0.5. An example of how these noninvariant data were simulated appears in Table 2 for the 2 observed groups, 2 latent classes, 20% noninvariant condition, for each degree of measurement noninvariance (i.e. 0, 0.1, 0.3, and 0.5). Consider as an example the noninvariance = 0 magnitude case, which is represented in the top section of Table 2. For all conditions simulated here, the target parameter is the probability of endorsing item 1, which appears in bold. Here, the latent class endorsement probabilities were identical for the two observed groups across the 5 items, as seen in Table

2. The noninvariance = 0.1 case with 20% noninvariant items appears in the second section of Table 2. In this case, latent class 1 in observed group 1 was simulated to have a proportion endorsing the dichotomous item of 0.8, while latent class 1 in observed group 2 was simulated to endorse the item at a rate of 0.7. Similarly, as seen in the third section of Table 2, for noninvariance = 0.3, latent class 1 in observed group 1 was simulated to have an endorsement probability for item 1 of 0.8, while latent class 1 in observed group 2 was simulated to have an endorsement probability of 0.5. A similar design was used for all other conditions included in this study.

2.7. Magnitude of Class Prevalence Inequality

In terms of latent class prevalence equivalence assessment, group 1 always had an equal distribution of latent class membership; i.e. 0.5/0.5 for 2 classes and 0.25/0.25/0.25/0.25 for 4 classes. Latent class prevalence rates then differed from group 2 by 0.05, 0.1, 0.15 or 0.3. Thus, for example, in the 0.3 prevalence difference condition, latent class 1 in group 1 was simulated to appear in 50% (0.5) of cases, while for group 2 latent class 1 was simulated to appear in 80% (0.8) of cases.

2.8. Study Outcome Variables

The outcome variables of interest were the Type I error rate (proportion of times noninvariance was indicated by a statistical procedure when data were simulated to be invariant in the population), and power (proportion of times noninvariance was correctly indicated by a statistical procedure when noninvariance was simulated in the data). The nominal Type I error rate set in this study is 0.05. It should be noted that with regard to the information indices, Type I error rate is a misnomer in that these statistics are not hypothesis tests, but rather methods of assessing relative model fit. However, for the sake of ease in reporting results consistently for all of the statistics studied here, this phrase will be applied to all of the methods included in this study. The use of information indices here was in keeping with standard practice, such that they were calculated for the constrained and unconstrained models and then compared with one another. The model with the lower value was taken to be the best fitting in the sample, and this result was compared with the actual state of affairs in the population. If the unconstrained (noninvariant) model was identified as better fitting than the constrained (invariant) model, then the conclusion based on the sample was that the latent class solutions were not invariant across the two known groups. When this decision was in error (i.e. the data were simulated to be invariant), the result was termed a Type I error, though in fact no hypothesis testing was conducted. However, in order to keep reporting of the results as clear and straightforward as possible, the terms Type I error rate and power will be used to signify incorrect and correct determinations of noninvariance. All such invariance testing was done under the assumption that the latent class models were properly specified for each of the observed groups.

Table 2. Example of measurement noninvariance simulation probabilities used to simulate data, by degree of measurement invariance, for the two group, 20% noninvariant condition, 2 latent classes condition.

	Noninvariance = 0			
	Observed group 1		Observed group 2	
Item	Latent class 1	Latent class 2	Latent class 1	Latent class 2
1	**0.8**	**0.2**	**0.8**	**0.2***
2	0.7	0.3	0.7	0.3
3	0.6	0.4	0.6	0.4
4	0.3	0.7	0.3	0.7
5	0.2	0.8	0.2	0.8
	Noninvariance = 0.1			
	Observed group 1		Observed group 2	
Item	Latent class 1	Latent class 2	Latent class 1	Latent class 2
1	**0.8**	**0.2**	**0.7**	**0.3***
2	0.7	0.3	0.7	0.3
3	0.6	0.4	0.6	0.4
4	0.3	0.7	0.3	0.7
5	0.2	0.8	0.2	0.8
	Noninvariance = 0.3			
	Observed group 1		Observed group 2	
Item	Latent class 1	Latent class 2	Latent class 1	Latent class 2
1	**0.8**	**0.2**	**0.5**	**0.5***
2	0.7	0.3	0.7	0.3
3	0.6	0.4	0.6	0.4
4	0.3	0.7	0.3	0.7
5	0.2	0.8	0.2	0.8
	Noninvariance = 0.5			
	Observed group 1		Observed group 2	
Item	Latent class 1	Latent class 2	Latent class 1	Latent class 2
1	**0.8**	**0.2**	**0.3**	**0.7***
2	0.7	0.3	0.7	0.3
3	0.6	0.4	0.6	0.4
4	0.3	0.7	0.3	0.7
5	0.2	0.8	0.2	0.8

*Parameter that is simulated to be noninvariant is in bold.

2.9. Analysis of Study Outcomes

In order to determine which of the manipulated conditions or their interactions were significantly related to the Type I error and power rates, repeated measures analysis of variance (ANOVA) was used where the rates across all 1000 replications served as the dependent variable, the statistic used to assess invariance was the within subjects factor, and the following manipulated conditions were between subjects factors.

3. RESULTS

The results presented below are divided into evaluations of Type I error and power rates for measurement invariance, and prevalence equivalence, respectively. For measurement invariance, the key points of interest were the comparative Type I error and power rates of the various methods across the levels of noninvariance, and whether any differences in these rates across the invariance testing methods interacted with the other manipulated conditions. Similarly, with respect to assessing the relative prevalence of the two latent classes, primary interest was on comparing the Type I error and power rates of the invariance testing methods, and how those rates changed with increasing differences in the prevalences of the latent classes across the observed groups. In addition, it was of interest to ascertain whether any such differences between the statistics' performance interacted with other manipulated factors in the study.

3.1. Measurement Invariance: Type I Error Rate

For the ANOVA for measurement invariance Type I error rate, the outcome was the rate of significant group difference findings for each replication when no group differences were simulated. The independent variables were method of invariance testing (method), number of latent classes (classes), class size ratio (classratio), group size ratio (group ratio), and sample size (N). The results of these analyses indicated that the only significant term was the interaction of method by classes by N $(p = 0.005, \eta^2 = 0.438)$. Figure 1 includes Type I error rates for method by classes by N. Of primary interest were the Type I error rates of the methods, for which, across conditions, the BIC, CAIC and aBIC demonstrated rejection rates of 0, except for aBIC with 2 classes and N of 200. In addition, AIC never had rejection rates greater than 0.05 for any of the simulated conditions. On the other hand, the G^2_Δ and BLRT tests had rates between 0.05 and 0.075 for 2 classes, whereas with 4 classes and Nof 2000 G^2_Δ had an inflated rate of 0.13, while BLRT maintained an error rate below 0.05.

3.2. Measurement Invariance: Power

For measurement invariance power, the 3-way interaction of method by N by magnitude of noninvariance (magnitude) was statistically significant $\left(p<0.001, \eta^2=0.232\right)$, as was the 3-way interaction of method by classes by group ratio $\left(p<0.001, \eta^2=0.139\right)$. Table 3 contains the power rates for the 6 methods by magnitude of difference and N. When interpreting these results, it is important to note that the G^2_Δ test was found to have inflated Type I error rates for $N=4$ and classes = 2000, as reported above. In terms of power, of primary interest were the relative performances of the various statistics for assessing invariance. Among these, BLRT displayed the highest power rates across conditions, while the G^2_Δ test consistently yielded the second highest power. The BIC, CAIC, and aBIC all had power values much lower than the other three methods across all conditions, with rates at or near 0 when the groups differed by 0.1.

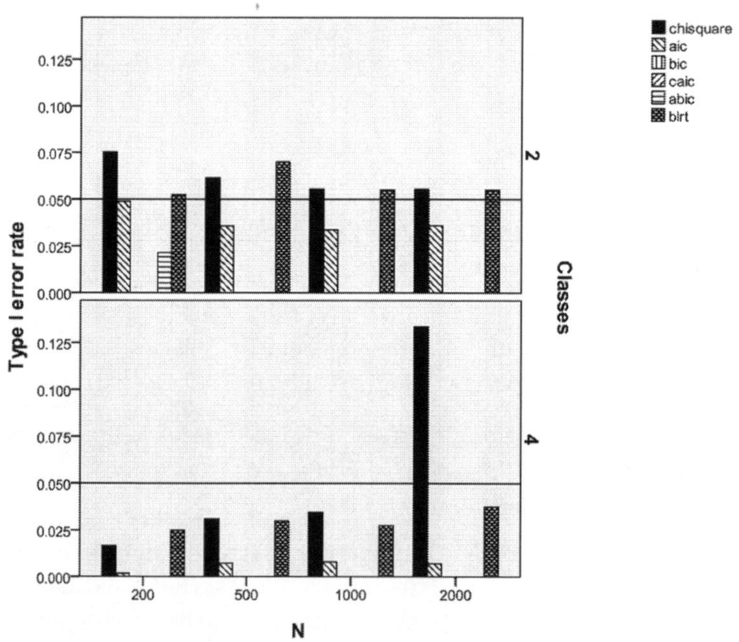

Figure 1. Type I error rate for measurement invariance testing by testing method, number of classes and sample size (N).

Table 3. Power for measurement invariance testing by testing method, magnitude of difference between groups (D), and sample size (N).

N	D	G^2_Δ	AIC	BIC	CAIC	aBIC	BLRT
	0.1	0.0770	0.0468	0.0000	0.0000	0.0211	0.0863
200	0.3	0.3120	0.2590	0.0020	0.0003	0.1955	0.3531
	0.5	0.4965	0.4564	0.0743	0.0306	0.4267	0.5362
	0.1	0.1472	0.0997	0.0000	0.0000	0.0004	0.2037
500	0.3	0.5305	0.4811	0.0447	0.0248	0.1623	0.5588
	0.5	0.6623	0.5626	0.3977	0.3205	0.4849	0.7169
	0.1	0.2788	0.2072	0.0000	0.0000	0.0007	0.3675
1000	0.3	0.6455	0.5488	0.2564	0.1852	0.3401	0.6456
	0.5	0.8474	0.7433	0.4983	0.4949	0.4996	0.8894
	0.1	0.4963	0.3813	0.0001	0.0000	0.0003	0.5650
2000	0.3	0.8064	0.6866	0.4193	0.3610	0.4314	0.8481
	0.5	0.9157	0.8833	0.4977	0.4943	0.4980	0.9631

Table 4. Power for measurement invariance testing by method, number of latent classes, and group ratio.

Classes	Group ratio	G^2_Δ	AIC	BIC	CAIC	aBIC	BLRT
2	1/1	0.7780	0.7510	0.3950	0.3516	0.5333	0.7821
	3/1	0.7386	0.7055	0.3351	0.2858	0.4866	0.7531
4	1/1	0.3567	0.2275	0.0002	0.0000	0.0003	0.5935
	3/1	0.1988	0.1014	0.0000	0.0000	0.0003	0.5547

Table 4 includes the power rates for method by classes by group ratio, the other significant interaction identified using ANOVA. An examination of the results reveals that for all of the methods power with 2 groups was slightly higher when the group ratios were equal, and the BLRT, G^2_Δ test and AIC had higher rates than the other approaches. When 4 classes were present, all of the methods displayed lower power than for 2 classes. In addition, the BLRT had higher power rates than any of the other methods, and both G^2_Δ and AIC demonstrated approximately twice as much power when the groups were of equal size as when they were unequal. Finally, BIC, CAIC and aBIC all had power rates at or near 0 for 4 latent classes.

3.3. Measurement Invariance Summary

To summarize the measurement invariance results presented above, all of the methods studied here maintained the nominal Type I error rate of 0.05, except for the G^2_Δ test with a sample size of 2000 and 4 latent classes present in the data. In terms of power, the BLRT yielded the highest rates across conditions, followed by the G^2_Δ test and AIC. The BIC, CAIC, and aBIC all yielded much lower power than the other three methods. Furthermore, all of the methods studied here yielded higher power rates when 2 latent classes were present, as opposed to 4. Thus, the BLRT can be said to yield the highest power among the methods studied here, while simultaneously maintaining the Type I error rate at approximately the nominal 0.05 level.

3.4. Equivalence of Latent Class Prevalences: Type I Error Rate

ANOVA results for the Type I error rate for assessment of latent class prevalence equivalence across observed groups found that the only significant term was the interaction of method by group ratio $\left(p = 0.0019, \eta^2 = 0.132\right)$. Results in Table 5 reveal that for all of the methods, Type I error was larger when the group ratio was 3/1 as opposed to equal. In addition, except for BLRT and aBIC, all of the methods had rates above 0.06 in the unequal group size condition, while they all had rates of approximately 0.055 or lower for equal group sizes, with the exception of AIC.

3.5. Equivalence of Latent Class Prevalences: Power

With respect to power for detecting a lack of latent class prevalence equivalence across groups, ANOVA identified two significant interactions: method by N by magnitude of noninvariance $\left(p < 0.001, \eta^2 = 0.633\right)$ and method by group ratio by magnitude of noninvariance $\left(p < 0.001, \eta^2 = 0.942\right)$. An examination of Table 6 reveals that across simulated conditions, BLRT displayed the highest power rates except for the greatest magnitude of group separation and N of 200. While AIC generally had the second highest power rates across conditions, it must be noted that it also displayed inflated Type I error rates across all conditions simulated here, as discussed previously. Therefore, these results must be interpreted with caution. Among the other methods, the G^2_Δ test displayed the highest power rates, again except for magnitude of noninvariance = 0.3 and N of 500 or more, in which case all of the statistics had power of 1.00. Finally, for $N = 200$, power never reached 0.8 for any of these techniques, except when the magnitude of difference was 0.3. For the lowest level of separation (0.05) the advantage of BLRT over the other methods was especially notable, particularly given that this method was also able to maintain the nominal Type I error rate of 0.05.

Table 7 contains power rates by method, group ratio and difference. For lower values of difference, power rates of BLRT were greater than those of the other statistics. When magnitude = 0.3, power rates for all of the methods were just below 1.00. Thus, when the group prevalence rates differed by as much as 0.3, all of the statistics studied here were able to detect the noninvariance. Again, however, both G^2_Δ and especially AIC had inflated Type I error rates for testing prevalence equality under many conditions, while BLRT did not.

Table 5. Type I error rate for testing equivalence of latent classes across groups, by method, class ratio, and group ratio.

Group ratio	G^2_Δ	AIC	BIC	CAIC	aBIC	BLRT
1/1	0.0545	0.1610	0.0135	0.0080	0.0185	0.0475
3/1	0.1863	0.3395	0.0905	0.0675	0.0540	0.0550

Table 6. Power for testing prevalence equivalence of latent classes across groups, by testing method, magnitude of difference between groups (D), and sample size (N).

D	*N*	G^2_Δ	AIC	BIC	CAIC	aBIC	BLRT
0.05	200	0.1720	0.3160	0.1075	0.0825	0.1915	0.4200
	500	0.2665	0.4410	0.1300	0.1040	0.0190	0.7500
	1000	0.3260	0.5085	0.1580	0.1270	0.0070	0.7450
	2000	0.5050	0.6880	0.2635	0.2235	0.0050	0.8800
0.10	200	0.3080	0.4725	0.2180	0.1675	0.3270	0.4500
	500	0.6560	0.7975	0.4655	0.3995	0.0800	0.8600
	1000	0.7675	0.8860	0.5495	0.4940	0.0750	0.9450
	2000	0.9485	0.9770	0.8445	0.8080	0.1225	0.9350
0.15	200	0.4635	0.6405	0.3540	0.2970	0.4830	0.5000
	500	0.9020	0.9560	0.8065	0.7615	0.2585	0.9550
	1000	0.9755	0.9920	0.9115	0.8760	0.3400	0.9750
	2000	1.0000	1.0000	0.9975	0.9945	0.6140	1.0000
0.30	200	0.9475	0.9845	0.9100	0.8740	0.9570	0.9750
	500	0.9995	0.9995	0.9995	0.9995	0.9855	0.9950
	1000	1.0000	1.0000	1.0000	1.0000	0.9990	1.0000
	2000	1.0000	1.0000	1.0000	1.0000	1.0000	1.0000

Table 7. Power for testing prevalence equivalence of latent classes across groups, by testing method, magnitude of difference between groups (D), and group size ratio.

D	Group ratio	G_s^2	AIC	BIC	CAIC	aBIC	BLRT
0.05	1/1	0.2472	0.4367	0.0927	0.0700	0.0250	0.7775
	3/1	0.3875	0.5400	0.2367	0.1985	0.0862	0.6200
0.10	1/1	0.6492	0.7810	0.4687	0.4092	0.0767	0.7900
	3/1	0.6907	0.7855	0.5700	0.5252	0.2255	0.8050
0.15	1/1	0.8375	0.9032	0.7652	0.7285	0.3040	0.8375
	3/1	0.8330	0.8910	0.7695	0.7360	0.5437	0.8775
0.30	1/1	0.9875	0.9962	0.9772	0.9670	0.9895	0.9725
	3/1	0.9860	0.9957	0.9775	0.9697	0.9812	0.9125

3.6. Summary of Prevalence Equivalence Testing Results

The BLRT method of assessing the equivalence of latent class prevalences across the observed groups was able to maintain the nominal 0.05 Type I error rate across all conditions simulated here, while also yielding the highest power rates. Therefore, much as with measurement equivalence, it appears that BLRT is able to both maintain the Type I error rate and yield relatively high power when testing for the equivalence of latent class prevalence across observed groups.

3.7. Analysis of Youth Risk Behavior Survey Data

In this section of the manuscript an LCA invariance analysis for data from the Youth Risk Behavior Survey (YRBS) is conducted to compare latent class solutions across genders. The YRBS is a biennial survey administered by the United States Centers for Disease Control and Prevention (CDC) designed for high school students. It consists of self-reports of past and current thoughts, emotions, behaviors, and exposure to health relevant curriculum within schools, and provides a snapshot of the general health and risky behavior of America's youth. The YRBS dataset is publicly available with individual identifying information, as well as school, state, and regional identifiers removed.

Of particular interest in this analysis is the assessment of gender differences in reckless behavior typologies among adolescents. Prior research has demonstrated that reckless physical behaviors increase in adolescence [28] . For example, as individuals move into their teen years the likelihood of their riding with an intoxicated driver increases [29] , while the likelihood of their regularly using their seatbelt in a car [30] , or wearing a helmet when riding a motorcycle decreases [31] . Furthermore, self care such as using sunscreen while outdoors [32] , or wearing a helmet when riding a bicycle [33] decline during this period as well. Finally, fighting has also been shown to increase in frequency and severity as individuals move from childhood to adolescence [34] . Coincident with this research demonstrating the increasing problem of reckless behavior in youth, there is also research to demonstrate that such risk taking may not be equally prevalent among males and females, and that in fact there may be some differences in the types of reckless behavior that are engaged in by the two genders [35] [36] .

Given this prior research, the goal of the current analysis was to investigate typologies of reckless behavior in the adolescent population using items from the YRBS, and to determine the extent to which these typologies were invariant for males and females. To this end, confirmatory LCA was used to ascertain whether the same number of latent classes was present for males and females, and if so whether there was measurement invariance for the items used to identify these latent classes, and if measurement invariance held, whether the prevalence of the latent classes was equivalent for the two genders. In short, the goal was to determine whether and to what extent reckless behavior typologies in the population are equivalent across males and females both in terms of the types of reported behaviors, and the relative frequency of the typologies in the population.

For this analysis, the items appearing in Table 8 (each coded as yes (1) or no (0)) were included in the inva- riance analysis. Each was selected because it measured a specific behavior that has been found in previous literature, cited above, to be associated with increased risk taking among adolescents. Of interest was whether the latent class solutions identified using these items were invariant for male and female respondents, first in terms of the number of latent classes found, second in terms of item response probabilities and if so, then third in terms of latent class prevalence. Based on the literature described above, it was anticipated that there would be distinct typologies of reckless behavior among adolescents, and that these patterns would differ by gender such that females would be relatively more risk averse than males.

Table 8. Latent class prevalence and response probabilities with 3 classes for the full sample.

Variable	Overall	Class 1 (0.25)	Class 2 (0.29)	Class 3 (0.46)
Rarely/Never wear bicycle helmet	0.88	0.94	0.65	1.00
Rarely/Never wear seat belt	0.10	0.27	0.01	0.07
Rode 1+ times with drinking driver	0.30	0.70	0.10	0.21
Fought 1+ times in last year	0.34	0.70	0.15	0.27
Rarely/Never wear motorcycle helmet	0.38	0.45	0.01	0.57
Rarely/Never wear sunscreen	0.92	0.95	0.80	0.97

A sample of 2000 subjects was drawn from the original YRBS data base, which contained 16,000 individuals. This sample of 2000 individuals was randomly drawn from among those who provided a valid response to each of the items appearing in Table 8 (a total of 13,743 respondents). In terms of gender, 1027 (51.9%) of the 2000 adolescents used in the LCA invariance analysis were male, as compared to 52.0% of those in the total sample. In order to cross-validate the results for this sample, we drew a second sample (without replacement) of 2000 individuals from the 11,743 remaining members of the YRBS data base who had responded to each of the items of interest, and conducted all of the analyses described below. While there were small differences in the percentages of respondents endorsing each item, the overall results (i.e. number of latent classes, noninvariance status) were the same for the second sample as for the first. This great similarity in the results for the two samples serves to support the results and conclusions described below.

Following the techniques for LCA invariance testing described in Collins and Lanza [5] , the first step was to identify the number of latent classes present in the sample as a whole. Four model solutions were fit to the YRBS data, for 2, 3, 4, and 5 latent classes. In order to determine the optimal number of latent classes, the fit indices described above, as well as model identification and interpretability of the solutions with regard to substantive meaning of the classes were considered. Based on the BLRT and the information indices, which appear in Table 9, particularly the AIC and aBIC, 3 latent classes appeared to fit the data best. Among the information indices, the aBIC in particular has been found to be most accurate in identifying the number of latent classes across a number of simulated conditions [14] [37] . In addition, research has found that among all currently available methods, the BLRT may be the most accurate approach for identifying the number of latent classes [6] . Therefore, given their generally greater accuracy, and the fact that they both (along with the AIC) identified the 3-class solution as optimal, this was the one selected here. In addition, a review of the nature of the latent classes, in terms of their item response profiles, suggested that the 3 latent class solution made substantive sense as well.

Table 8 includes the prevalence of each latent class, along with the proportion of individuals in the classes that endorsed each of the items used in the analysis, for the total sample. Based on these results, it appears that latent class 1 can be described as relatively high risk compared to the other two. Individuals in this group, which made up 25% of the total sample, were the least likely to use transportation safety devices such as seat belts in cars and helmets on bicycle/motorcycles, and were most likely to ride with a drinking driver and engage in fighting in the last year. Latent class 2 was the most risk averse, being more likely to wear helmets on bicycles and motorcycles, more likely to wear a seat belt, and to use sunscreen, and the least likely to have ridden with a drinking driver or to have fought. Finally, latent class 3 displayed a pattern of behaviors that put them in between the highest and lowest risk groups.

Table 9. Model fit indices for LCA models with the full sample and with each gender.

			Total sample			
Latent classes	G^2 (df)	AIC	BIC	CAIC	aBIC	BLRT *p*-value
2	96.07 (50)	122.07	194.88	207.88	153.58	0.001
3	57.05 (43)	93.34	209.06	229.06	145.52	0.901
4	40.34 (36)	97.04	245.56	272.56	159.78	0.876
5	30.34 (39)	98.34	288.77	322.77	180.75	0.922
			Females			
Latent classes	G^2 (df)	AIC	BIC	CAIC	aBIC	BLRT *p*-value
2	61.14 (50)	88.01	150.75	163.75	122.35	<0.001
3	48.01 (43)	87.14	188.87	205.35	109.46	0.863
4	43.54 (36)	97.54	229.64	256.64	143.89	0.856
5	31.02 (39)	97.99	292.35	319.09	179.80	0.738
			Males			
Latent classes	G^2 (df)	AIC	BIC	CAIC	aBIC	BLRT *p*-value
2	70.03 (50)	96.03	159.91	172.91	118.62	<0.001
3	42.86 (43)	82.86	181.13	201.13	117.61	0.911
4	30.55 (36)	84.55	217.22	244.22	131.47	0.917
5	28.74 (39)	89.01	300.46	318.17	155.66	0.936

After determining the number of latent classes for the sample as a whole, the next step was to establish that the number of latent classes and their general patterns of item responses were similar to those of the overall population. In other words, in order to test whether the item response probabilities for the two genders were invariant, we must first ensure that the number of latent classes in each was the same. LCA was applied to each gender individually, and as can be seen inTable 9, for both males and females the 3 class solution proved optimal, based on the AIC and aBIC. An examination of Table 10 shows that for each gender, latent class 1 reported relatively high rates of risky behavior, while latent class 2 reported the least risky behaviors and class 3 fell somewhere in between the two. Thus, it does appear that the basic constituencies of the latent classes were present for both males and females. Before we can talk about any gender differences in the prevalence of these behaviors for specific latent classes, we must first assess whether LCA invariance holds or not.

As described above, in order to determine whether measurement invariance holds, we fit two models. In the first model all item response probabilities were constrained to be equal across the genders, while in the second model they were all unconstrained. If the difference in model fit favored the unconstrained model (or was statistically significant in the case of the G^2_Δ test) we would conclude that invariance did not hold. For both the constrained and unconstrained models, we fit the 3 class solution, given that it had already been established as optimal for both the full and gender specific samples. The fit statistics for the constrained and unconstrained models appear in Table 11, along with the G^2_Δ test and BLRT. BLRT, the G^2_Δ test, and AIC indicated that the

Table 10. Latent class prevalence and response probabilities with 3 classes by gender.

	Females ($N = 944$)		
Variable	Class 1 (0.28)	Class 2 (0.33)	Class 3 (0.40)
Rarely/Never wear bicycle helmet	1.00	0.60	1.00
Rarely/Never wear seat belt	0.16	0.01	0.03
Rode 1+ times with drinking driver	0.51	0.11	0.23
Fought 1+ times in last year	0.54	0.11	0.01
Rarely/Never wear motorcycle helmet	0.46	0.06	0.43
Mostly/Always wear sunscreen	0.05	0.28	0.01
	Males ($N = 1027$)		
Variable	Class 1 (0.25)	Class 2 (0.39)	Class 3 (0.36)
Rarely/Never wear bicycle helmet	0.93	0.78	0.99
Rarely/Never wear seat belt	0.31	0.03	0.10
Rode 1+ times with drinking driver	0.81	0.11	0.13
Fought 1+ times in last year	0.78	0.25	0.38
Rarely/Never wear motorcycle helmet	0.46	0.00	0.80
Mostly/Always wear sunscreen	0.04	0.11	0.02

Table 11. Fit statistics and G^2 difference test for the invariant and noninvariant models.

Model	G^2 (df)	AIC	BIC	CAIC	aBIC	BLRT p-value
Constrained	153.94 (105)	197.94	321.06	343.06	257.16	
Unconstrained	90.84 (87)	170.84	394.70	434.70	267.62	
Difference	63.1 (18)*					<0.001

*$p = 0.00000064$.

noninvariant LCA model provided better fit to the data, while the BIC, CAIC and aBIC all suggested that the best fit was given by the constrained model. Based on the results of the simulation study described previously, it would appear that the BLRT, in particular, may provide more accurate results for measurement invariance than the other approaches, because it displays higher power while maintaining the nominal (0.05) Type I error rate. Thus we will work under the presumption that item response probabilities did differ between clusters across genders. An examination of Table 10 suggests that females in the highest risk group (latent class 1) were more likely to wear a safety belt, and less likely to fight or ride with a drinking driver than their male counterparts. Similarly, females in the lowest risk category (latent class 2), were less likely to engage in risky behaviors such as not wearing a bicycle helmet, not wearing sunscreen or fighting than were males in the low risk class. Indeed, the same pattern across genders was also seen in latent class 3, where females were less likely to engage in risky behaviors (i.e. not wearing a seat belt, fighting, not wearing a motorcycle helmet) than males, except for riding with a drinking driver, which they were actually more likely to do than were males in the corresponding latent class.

These results appear to support the a priori hypothesis that the nature of risk taking typologies differs between males and females. Male and female respondents did appear in the same basic latent classes based on their likelihood to engage in risky behaviors, including those who were relatively risk averse, those who were more likely to engage in reckless behavior, and a third group of individuals who were in the middle of the other two in terms of their likelihood to engage in reckless behavior. However, with respect to endorsement of specific reckless behaviors, females in each of these classes were more risk averse than their male counterparts, with the exception of latent class 3 and riding with a drinking driver. Thus, the hypothesis that there would be different typologies of individuals with respect to their propensity for engaging in reckless behaviors was upheld, as was the hypothesis that females would be less likely than males to engage in such behaviors regardless of typology. Given the lack of measurement invariance, it is not recommended that we test for the equality of latent class prevalences across groups [5] .

4. DISCUSSION

The goal of this study was to compare several statistics in terms of their ability to correctly ascertain whether group invariance was present in the context of MGLCA. In addition, the use of MGLCA to conduct an invariance analysis with risk assessment items from the YRBS was also demonstrated. It is hoped that the results presented here will aid researchers from a variety of fields in successfully conducting invariance testing with LCA models. The outcomes described above suggest that BLRT is the optimal method, of those studied here, for assessing the invariance question. Whether considering measurement invariance or the equivalence of class prevalences, the BLRT most consistently maintained the Type I error rate of 0.05, while simultaneously yielding among the highest power values in the simulation study. This combination of Type I error control with relatively high power was particularly notable in the comparison of class prevalence estimates across groups, where virtually all of the other methods either kept the error rate at or below the nominal 0.05 level and had low power (i.e. CAIC, aBIC), or had higher power but were unable to control the Type I error rate (i.e. G^2_Δ, AIC). Considering the hypotheses built upon earlier work and stated above, it does appear that BLRT is the most accurate method for investigating LCA invariance. This result is in keeping with work by Nylund, Asparouhov and Muthén [6] showing that BLRT was also optimal for selecting the number of latent classes. In addition, the G^2_Δ difference test did display some sensitivity to large sample size for assessing measurement invariance with 4 classes, but not in most other simulated conditions, thus partially matching earlier work showing its sensitivity to large N [20] . Additionally, as noted above, prior research in the LCA model selection literature found that AIC had a tendency to identify the presence of more latent classes in the sample than were actually present in the population [15] . This tendency to be overly liberal was evident for the AIC in comparing the equivalence of group prevalence estimates as well, where it too often indicated that the groups differed in this regard when they did not. Finally, consistent with the mixed results from the body of prior research with regard to information indices, no one of them was found to be clearly superior to the others. While the CAIC, BIC, and aBIC did not have the Type I error inflation of the AIC, neither did they have its power for rejecting the invariance null hypothesis when it was incorrect. For example, for determining whether the latent class specific item response probabilities differ across known groups (i.e. measurement invariance), the AIC and perhaps the G^2_Δ performed better than did CAIC, BIC, or aBIC in terms of power and Type I error. Indeed, while having error rates at or near 0, these latter three methods also had very low power for measurement invariance when compared to the other techniques. Therefore, it is not clear that having such low error rates is truly a positive characteristic, given the concomitant low power rates, as it appears that these statistics are simply unlikely to reject the invariant model whether it is true or not. Finally, it was of interest that the aBIC did not perform better than the other information indices, although in some model selection studies it had been shown to be superior [13] [14] . Clearly, further investigation of this statistic in the context of confirmatory LCA is called for.

In addition to providing insights into which statistics worked best under a variety of conditions, this study also revealed situations in which invariance testing in LCA might be somewhat more problematic. Specifically, when the observed groups differ in size by as much as a 3/1 ratio, several of the statistical approaches to assessing LCA invariance will be adversely affected, yielding lower power, particularly in the presence of more latent classes. On the other hand, the null hypothesis of equivalence of latent class prevalences will too frequently be rejected incorrectly in the unequal known group sample size case, with the exception of aBIC and BLRT. While the BLRT did have slightly lower power values in the 3/1 group ratio case, this difference was quite modest when compared to the other methods, typically decreasing by less than 0.05. In addition, study results show that when more latent classes are present in the population, the ability to accurately detect a lack of invariance may be compromised. Indeed, across most of the methods studied here and most of the simulated conditions, power for finding such differences declined from the 2 class condition. This decline was much less precipitous, however, for the BLRT when compared to the other methods.

4.1. Limitations and Directions for Future Research

By its very nature, simulation research presents some limitations in that the results are only generalizable to real world situations that mirror those simulated. While the conditions used here were selected in order to match real world examples in the literature, they could not be exhaustive. Therefore, future research in this area needs to continue in several directions. First of all, the observed variables simulated here were all dichotomous in nature. Thus, future studies should focus on the application of MGLCA to the case of polytomous data. In addition, the use of 5 indicators, while relatively common in practice, was not varied, so that subsequent work needs to expand the conditions for the number indicators so as to ascertain to what extent it may impact both types of invariance testing. The number of latent classes was also held constant for the two groups, which may not reflect all conditions encountered in actual practice. Therefore, future research should also consider cases when the number of classes in fact differs across observed groups. Likewise the number of groups was held constant at 2. While this is also very common in actual practice, it is nonetheless a limitation of this study in that we didn't learn how well MGLCA can correctly determine invariance when the number of groups is larger. Finally, future studies should also examine a broader range of group size ratios. These results suggest strongly that the relative sizes of the groups matters, particularly in terms of testing the relative prevalence of latent classes. The two conditions simulated here, 1/1 and 3/1, were selected to represent two disparate and perhaps extreme cases. However, it would be very informative for practitioners wishing to use MGLCA to know at what point between the 1/1 and 3/1 ratios group size inequality becomes problematic.

4.2. Recommendations for Practice

Holmes Finch (6,63)">6] , the BLRT may be the optimal tool for assessing invariance in the context of LCA. It was able to both maintain the nominal Type I error rate across virtually all simulated conditions, and yielded power that was always as high as, or higher than the other methods. Furthermore, the BLRT worked relatively well for both types of invariance assessment, which was not the case for the other statistics studied here. For this reason, the BLRT appears to be an optimal choice. In addition, researchers must always consider the structure of their data when interpreting MGLCA results. The presence of relatively many latent classes, coupled with unequal observed group sizes will result in diminished power, even for BLRT. Thus, in such conditions, those engaged in invariance assessment must carefully consider the meaning of a non-significant test result. Nonetheless, the BLRT is clearly a promising method for assessing LCA invariance and as such should prove to be quite useful for researchers interested in using MGLCA to conduct this analysis.

REFERENCES

1. Hoijtink, H. (2001) Confirmatory Latent Class Analysis: Model Selection Using Bayes Factors and (Pseudo) Likelihood Ratio Statistics. Multivariate Behavioral Research, 36, 563-588.

2. McCutcheon, A.L. (2002) Basic Concepts and Procedures in Single- and Multiple-Group Latent Class Analysis. In: Hagenaars, J.A. and McCutcheon, A.L., Eds., Applied Latent Class Analysis, Cambridge University Press, Cambridge, 56-86.

3. Laudy, O., Boom, J. and Hoijtink, H. (2005) Bayesian Computational Methods for Inequality Constrained Latent Class Analysis. In: van der Ark, L.A., Croon, M.A. and Sijtsma, K., Eds., New Developments in Categorical Data Analysis for the Social and Behavioral Sciences, Lawrence Erlbaum Associates Publishers, Mahwah, 63-82.

4. Finch, W.H. and Bronk, K.C. (2011) Conducting Confirmatory Latent Class Analysis Using Mplus. Structural Equation Modeling, 18, 132-151.

5. Collins, L.M. and Lanza, S.T. (2010) Latent Class and Latent Transition Analysis: With Applications in the Social, Behavioral and Health Sciences. Wiley, New York.

6. Nylund, K.L., Asparouhov, T. and Muthén, B.O. (2007) Deciding on the Number of Classes in Latent Class Analysis and Growth Mixture Modeling: A Monte Carlo Simulation Study. Structural Equation Modeling, 14, 535-569.

7. French, B.F. and Finch, W.H. (2006) Confirmatory Factor Analytic Procedures for the Determination of Measurement Invariance. Structural Equation Modeling, 13, 378-402.

8. Lubke, G.H. and Muthén, B.O. (2004) Applying Multigroup Confirmatory Factor Models for Continuous Outcomes to Likert Scale

Data Complicates Meaningful Group Comparisons. Structural Equation Modeling, 11, 514-534.

9. Cheung, G.W. and Rensvold, R.B. (2002) Evaluating Goodness-of-Fit Indexes for Testing Measurement Invariance. Structural Equation Modeling, 9, 233-255.

10. Akaike, H. (1987) Factor Analysis and AIC. Psychometrika, 52, 317-332.

11. Bozdogan, H. (1987) Model Selection and Akaike's Information Criterion (AIC): The General Theory and Its Analytical Extensions. Psychometrika, 52, 345-370.

12. McLachlan, G. and Peel, D. (2000) Finite Mixture Models. Wiley, New York.

13. Tofighi, D. and Enders, C.K. (2007) Identifying the Correct Number of Classes in a Growth Mixture Model. In: Hancock, G.R., Ed., Advances in Latent Variable Mixture Models, Information Age, Greenwich, 317-341.

14. Yang, C. (2006) Evaluating Latent Class Analyses in Qualitative Phenotype Identification. Computational Statistics & Data Analysis, 50, 1090-1104.

15. Celeux, G. and Soromenho, G. (1996) An Entropy Criterion for Assessing the Number of Clusters in a Mixture Model. Journal of Classification, 13, 195-212.

16. Jedidi, K., Jagpal, H. and DeSarbo, W.S. (1997) Finite-Mixture Structural Equation Models for Response-Based Segmentation and Unobserved Heterogeneity. Marketing Science, 16, 39-59.

17. Read, T.R.C. and Cressie, N.A.C. (1988) Goodness of Fit Statistics for Discrete Multivariate Data. Springer, New York.

18. Koehler, K.J. and Larntz, K. (1980) An Empirical Investigation of Goodness of Fit Statistics for Sparse Multinomials. Journal of the American Statistical Association, 75, 336-344.

19. Koehler, K.J. (1986) Goodness-of-Fit Tests for Log-Linear Models in Sparse Contingency Tables. Journal of the American Statistical Association, 81, 483-493.

20. Meade, A.W. and Lautenschlager, G.J. (2004) A Monte-Carlo Study of Confirmatory Factor Analytic Tests of Measurement Equivalence/Invariance. Structural Equation Modeling, 11, 60-72.

21. Burnham, K. and Anderson, D. (2003) Model Selection and Multimodel Inference: A Practical-Theoretic Approach. Springer-Verlag, New York.

22. Vrieze, S.I. (2012) Model Selection and Psychological Theory: A Discussion of the Differences between the Akaike Information Criterion (AIC) and the Bayesian Information Criterion (BIC). Psychological Methods, 17, 228-243.

23. Shao, J. (1997) An Asymptotic Theory for Linear Model Selection (with Discussion). Statistica Sinica, 7, 221-242.

24. Wicherts, J.M. and Dolan, C.V. (2004) A Cautionary Note on the Use of Information Fit Indexes in Covariance Structure Modeling with Mean. Structural Equation Modeling, 11, 45-50.

25. Dziak, J.J., Lanza, S.T. and Xu, S. (2011) SimulateLcaDataset SAS Macro Users Guide, Version 1.1.0. The Methodology Center, Pennsylvania State University, University Park.http://methodology.psu.edu

26. Lanza, S.T., Dziak, J.J., Huang, L., Xu, S. and Collins, L.M. (2011) PROC LCA & PROC LTA Users' Guide (Version 1.2.3). The Methodology Center, Pennsylvania State University, University Park. http://methodology.psu.edu

27. SAS Institute (2009) SAS Version 9.1. SAS Institute, Cary.

28. Arnett, J. (1995) The Young and the Reckless: Adolescent Reckless Behavior. Current Directions in Psychological Science, 4, 67-70.

29. Poulin, C., Boudreau, B. and Asbridge, M. (2007) Adolescent Passengers of Drunk Drivers: A Multi-Level Exploration into the Inequities of Risk and Safety. Addiction, 102, 51-61.

30. McCartt, A.T. and Northrup, V.S. (2004) Factors Related to Seat Belt Use among Fatally Injured Teenage Drivers. Journal of Safety Research, 35, 29-38.

31. Bianco, A., Trani, F., Santoro, G. and Angelillo, I.F. (2005) Adolescents' Attitudes and Behavior towards Motorcycle Helmet Use in Italy. European Journal of Pediatrics, 164, 207-211.

32. Everett, S., Miyamoto, J., Saraiya, M. and Berkowitz, Z. (2012) Trends in Sunscreen Use among U.S. High School Students: 1999-2009. The Journal of Adolescent Health, 50, 304-307.

33. Klein, K., Thompson, D., Scheidt, P., Overpeck, M. and Gross, L., HBSC International Investigators (2005) Factors Associated with Bicycle Helmet Use among Young Adolescents in a Multinational Sample. Injury Prevention, 11, 288-293.

34. Lowry, R., Powell, K.E., Kann, L., Collins, J.L. and Kolbe, L.J. (1998) Weapon-Carrying, Physical Fighting, and Fight Related Injury among

U.S. Adolescents. American Journal of Preventive Medicine, 14, 122-129.

35. Hirschberger, G., Florian, V., Mikulinger, M., Goldenberg, J.L. and Pyszczynski, T. (2002) Gender Differences in the Willingness to Engage in Risky Behavior: A Terror Management Perspective. Death Studies, 26, 117-141.

36. Byrnes, J.P., Miller, D.C. and Schafer, W.D. (1999) Gender Differences in Risk Taking: A Meta-Analysis. Psychological Bulletin, 125, 367-383.

37. Sclove, L. (1987) Application of Model-Selection Criteria to Some Problems in Multivariate Analysis. Psychometrika, 52, 333-343.

Chapter 7

Enhancing Air Quality Forecasts over Catalonia (Spain) Using Model Output Statistics

Víctor Andrés Pérez[1], Raúl Arasa[1*], Bernat Codina[1,2], Jesica Piñón[1]

[1]*Technical Department, Meteosim S.L., Barcelona, Spain*

[2]*Department of Astronomy and Meteorology, University of Barcelona, Barcelona, Spain*

ABSTRACT

Model Output Statistics (MOS) is a well-known technique that allows improving outputs from numerical atmospheric models. In this contribution, we present the development of a MOS algorithm to improve air quality forecasts in Catalonia, a region in the northeast of Spain. These forecasts are obtained from an Eulerian coupled air quality modelling system developed by Meteosim. Nitrogen Dioxide (NO_2), Particulate Matter (PM_{10}) and Ozone (0_3) have been the pollutants considered and the methodology has been applied on statistical values of these pollutants according to regulatory levels. Four MOS algorithms have been developed, characterized by different approaches in relation with seasonal stratification and stratification according to the measurement stations considered. Algorithms have been compared among them in order to obtain a MOS that reduces the forecast uncertainties. Results obtained show that the best MOS designed increases the accuracy of NO_2 maximum 1-h daily value forecast from 71% to 75%, from 68% to 81% in the case of daily values of PM_{10}, and finally, the accuracy of O_3 maximum 1-h daily value from 79% to 87%.

Keywords: Air Quality Modelling, Forecasting, Model Output Statistics (MOS)

1. INTRODUCTION

Air quality is one of the main issues that concerns current atmospheric research. Global air pollution has an impact on human health [1] , climate change [2] and on the physics and chemistry of the atmosphere [3] . Meteorological and air quality related environmental phenomena influence and limit the regional and urban development and safety management, which often leads to severe negative impacts on public health, economy and environment of polluted areas.

In Spain, annual average values of Nitrogen Dioxide (NO_2) and Particulate Matter (PM_{10}) are elevated in many urban air quality measurement stations with traffic influence [4] . Whereas, high ozone levels are measured in rural or suburban areas located downwind of urban or industrial locations and where local ozone precursors are lacking [5] [6] .The area of study is Catalonia (41.82˚N, 1.47˚E) in north-eastern Spain, in which six episodes with high levels of pollutants have been recorded in the last five years, thus making the Catalan Government adopt temporary mitigation measures. Figure 1 shows the topographical features of Catalonia. Black dots point the four provinces that divide the studied area.

For these reasons, it will be desirable that public administrations may count on tools which enable them to anticipate the potential risks caused by pollution, helping the management of health, economical and environmental impacts that may affect the polluted areas of its population. In this sense, models are a very useful tool for local administrations for planning and managing production, human resources, activities and emergency procedures; and to introduce improvement plans of air quality urban areas. Nowadays, coupled Eulerian air quality models are useful tools to manage air pollution, and they can even complement or replace air quality monitoring in terms of what the European Directive establishes.

Despite years of refinements and improvements, meteorological and air quality models still contain significant errors. Several statistical techniques have been designed and applied in the last few years [7] -[12] in order to reduce the uncertainty of their forecasts. In this work, we focus on the use of the Model Output Statistics (MOS) technique to improve forecasts obtained by an air quality modelling system. MOS has been applied usually to meteorological applications [13] -[17] but there are a few applications on air quality [18] -[20] . Air quality forecasts have been obtained by a coupled Eulerian air quality modelling system developed by Meteosim [21] . This coupled air quality modelling system has been applied and tested successfully in urban (Madrid or Barcelona in Spain; Nice in France) and industrial areas (Ponferrada or Tarragona in Spain). This air quality modelling system has been evaluated using Maximum Relative Directive Error [22] referred in the European

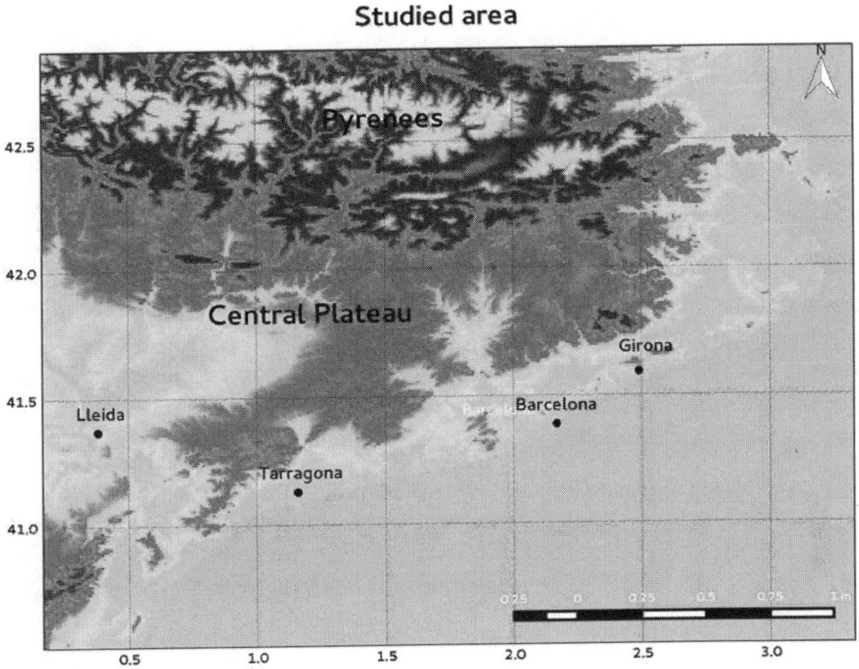

Figure 1. Topographical features of Catalonia.

Directive EC/2008/50. Results obtained from this evaluation accomplish the model uncertainty limits according to the Directive for the pollutants O_3, NO_2, PM_{10}, SO_2 and CO, having used measurements from more than 120 stations (urban, suburban and rural locations) during a period of three years.

The MOS technique has been applied to daily values of NO_2 and PM_{10}, maximum 1-h daily values of NO_2 and O_3 and maximum 8-h daily values of O_3 forecasted over Catalonia (Spain). Air quality measurement data have been provided by the Air Quality Network that belongs to the Territory and Sustainability Department of the Catalonian Government. The study includes a numerical deterministic evaluation that shows the accuracy of the air quality modelling outputs with and without the MOS technique.

Description of the model output statistics technique and the uncertainty evaluation methodology are presented in section 2; a detailed analysis of the results obtained is presented in section 3; and finally, some conclusions are reported in section 4.

2. METHODOLOGY

2.1. Model Output Statistics

The technique Model Output Statistics (MOS) computes regression equations between observed and model forecasted variables. These equations are later applied to the raw model output to get a modified and statistically corrected output.

This technique requires a set of variables from the modelling system to be corrected and a set of observed variables. These datasets cover a time period known as training period. Its length is variable but it is advisable to have the longest possible period to include potential cycles. In this work we have used the period comprised between the 24th of October 2012 and the 31st of August 2013, which includes 720,000 hourly values.

Within the MOS correlation equations we distinguish between predictands and predictors. The predictant is the dependent variable to be forecasted or corrected. In developing our MOS, the predictants will be the statistics described in the introduction: maximum 1-h value and daily values for NO_2, daily value for PM_{10} and maximum 1-h and 8-h values for O_3.

The predictors are the independent variables of the regression equations. In this work, we have used the concentration values forecasted by the modelling system. Since we have only used one predictor, we will omit the discussion about predictor-selection algorithms [23] -[27] . It is worth mentioning that in a preliminary phase we did test the viability of using more predictors, such as hourly data and its derived statistics, as well as past observational data. However, we finally found out that the use of the statistics of each pollutant as obtained by the modelling system produced the best results.

Often, the datasets employed in the regression equations are grouped in order to optimize the forecasting relationships and improve the precision of the MOS technique. This grouping is usually referred to as data stratification.

2.2. Error Statistics

In order to evaluate the potential improvements of the MOS technique on the modelling system, we will use a numerical validation. Several statistics will be used for the error measurement of the differences between results obtained with and without MOS. Most of them are typical statistics employed in error evaluation in meteorological models [22] [28] and, in particular, in air quality models [29] -[31] . Reference [32] draws some recommendations on the maximum thresholds for some of these statistics in the analysis and forecast of pollutant concentrations[1] that we will use in this work.

In the next definitions, n is the sample size, o_i and f_i the observed and forecasted values, where i Î [1, n]:

Mean normalized error:

$$\mathrm{MNBE} = \frac{1}{n}\sum_{i=1}^{i=n} \frac{f_i - o_i}{o_i} \times 100 \tag{1}$$

Reference [32] recommendation: MNBE < ±15%.

Mean normalized gross error:

$$\mathrm{MNGE} = \frac{1}{n}\sum_{i=1}^{i=n} \left| \frac{f_i - o_i}{o_i} \right| \times 100 \tag{2}$$

Reference [32] recommendation: MNGE < 35%.

Index of agreement:

$$\mathrm{IOA} = 1 - \frac{\sum_{i=1}^{i=n} |f_i - o_i|^2}{\sum_{i=1}^{i=n} |f_i - \bar{o}_i| + |o_i - \bar{o}_i|} \tag{3}$$

This is a statistic to evaluate with a single value the goodness of fit of a modelling system with respect to the observations. As close as one is the value, the best is the fit; and it worsens as it approaches to zero.

2.3. Forecast Evaluation

Table 1 shows a summary of the errors of the modelling system in the training period using the above described statistics.

As can be seen, most of the statistics fall within [32] recommendations, with the exception of the values of the PM_{10} pollutant. The presence of these errors, however, justifies the need of developing a MOS.

2.4. Description of the Proposed MOS Categories

From the observed and modelled data, we have proposed two kinds of stratifications:

- Seasonal stratification. The analysis of observed concentration exhibits a dependency on temperature, especially remarkable for ozone, which suggests splitting the training stage in 4 different periods. At first glance one could think that a natural division would involve climatologically seasons, but the study of mean temperature and concentration of the considered pollutants show that this division is not appropriate. We have finally chosen a division in training periods in which the values of temperature and pollutants change significantly from one to another (Figure 2 and Table 2).These periods turned to be 23 Oct 2012 to 02 March 2013 (defined as Period 1), 03 March 2012 to 01 May 2013 (defined as Period 2), 02 May 2013 to 06 July 2013 (defined as Period 3) and 07 July 2013 to 31 August 2013 (defined as Period 4).Stratification according to the measurement station in order to identify systematic errors and biases of each measurement point.

- Stratification according to the measurement station in order to identify systematic errors and biases of each measurement point.

Table 1. Summary of the errors of modelling system in the training period.

Pollutant	Description	MNBE	MNGE	IOA
NO_2	Maximum 1-h value	11.7%	29.5%	0.863
	Daily value	11.8%	24.9%	0.912
PM_{10}	Daily value	−24.5%	32.1%	0.751
O_3	Maximum 1-h value	2.5%	21.3%	0.618
	Maximum 8-h value	7.6%	25.9%	0.678

Table 2. Average maximum hourly and daily values of NO_2, daily value of PM_{10} and maximum hourly and 8-hourly values of O_3 within each period.

Pollutant	Statistic	Observed Mean (μgm^{-3})			
		Period 1	Period 2	Period 3	Period 4
NO_2	Maximum 1-h value	68	58	53	53
	Daily value	37	28	27	27
PM_{10}	Daily value	18	14	17	18
O_3	Maximum 1-h value	66	95	100	120
	Maximum 8-h value	59	87	91	108

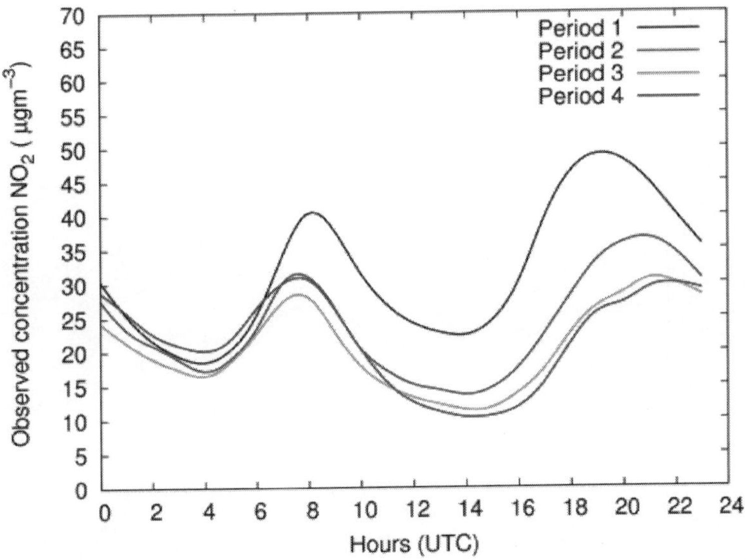
Period 1
Period 2
Period 3
Period 4
Observed concentration NO2 (µgm-3)
0
5
10
15
20
25
30
35
40
45
50
55
60
65
70
0 2 4 6 8 10 12 14 16 18 20 22 24
Hours (UTC)

(a)

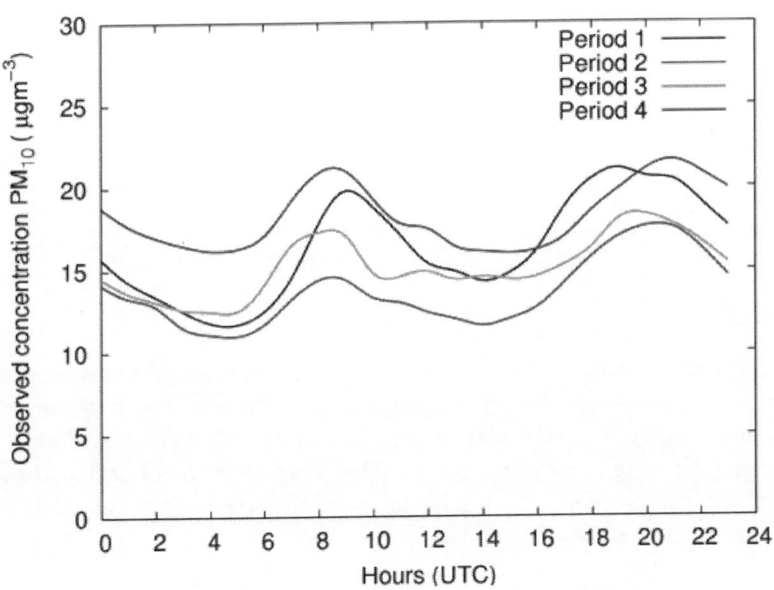
Period 1
Period 2
Period 3
Period 4
Observed concentration PM10 (µgm-3)
0
5
10
15
20
25
30
0 2 4 6 8 10 12 14 16 18 20 22 24
Hours (UTC)

(b)

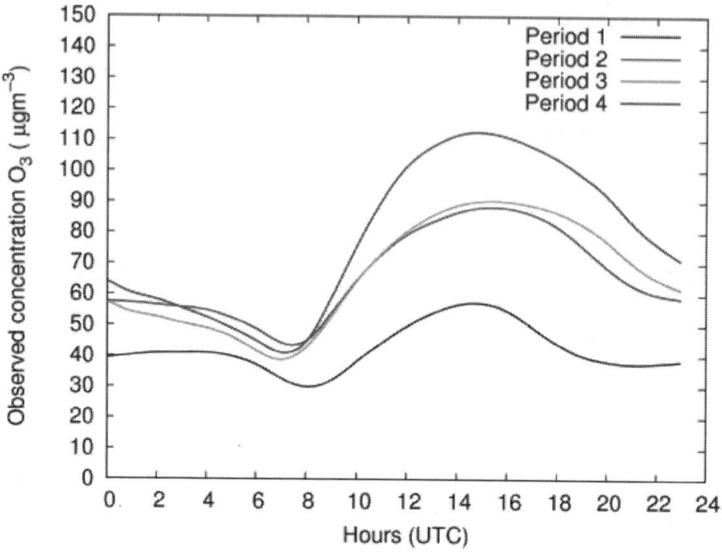

(c)

Figure 2. Average daily time evolution of concentration values of NO_2 (a) PM_{10} (b) and O_3(c) for the considered periods.

A total of 2 × 2 separations have been done, yielding to 4 different categories. Table 3 shows a summary of the categories and the nomenclature that will be used.

3. RESULTS

In this section we analyze the errors resulting from applying a period-specific trained MOS on the modelling output system. We use the deterministic numerical validation, the comparison between mean daily evolutions, and the comparison of the frequency histograms of the average errors of each category as well as the parameters of a Gaussian fit. This will allow us to choose the best MOS for each pollutant and statistics.

3.1. NO_2

Table 4 shows the error analysis statistics of the different categories after having been applied to the modelling system of the NO_2 pollutant. Figure 3 shows a comparison between normalized mean absolute errors in each category.

Table 4 and Figure 3 show that MOS E C produces the best results for both the maximum hourly value and the daily value. Notice that MOS 0 and MOS C do not improve the modelling system.

In case of the maximum hourly value, MOS E C reduces MNBE from 11.7% to 9.5% of the modelling system, reduces MNGE from 29.5% to 24.3%, and increases IOA from 0.86 to 0.91.

For the daily value, MOS E reduces MNBE from 11.8% to −0.9%, thus correcting the overestimation of the modelling system, reduces MNGE from 24.9% to 18.1%, and increases IOA from 0.91 to 0.95.

The frequency histograms of Figure 4(a) and Figure 4(b) have been created to better analyze the improvements of the different categories. These histograms display the error distributions of the modelling system and of maximum hourly values and daily values for each MOS category. Obviously, better results are associated to centred and peaked distributions.

Table 3. Summary of the proposed MOS categories.

	No stratification by measurement stations	With stratification by measurement stations
No stratification by periods	MOS 0	MOS E
With stratification by periods	MOS C	MOS E C

Table 4. Errors of all MOS categories as well as NO_2 forecasting system.

	Maximum 1-h value					Daily value				
	Observed mean ($\mu g \cdot m^{-3}$)	Modelled mean ($\mu g \cdot m^{-3}$)	MNBE (%)	MNGE (%)	IOA	Observed mean ($\mu g \cdot m^{-3}$)	Modelled mean ($\mu g \cdot m^{-3}$)	MNBE (%)	MNGE (%)	IOA
SIAM	62	65	11.7	29.5	0.863	33	35	11.8	24.9	0.912
MOS 0	62	62	20.4	37.3	0.847	33	33	1.1	32.0	0.913
MOS E	62	62	10.4	26.2	0.887	33	33	6.7	20.1	0.943
MOS C	62	62	−12.0	37.5	0.853	33	33	−4.1	30.6	0.917
MOS E C	62	62	9.5	24.3	0.905	33	33	−0.9	18.1	0.953

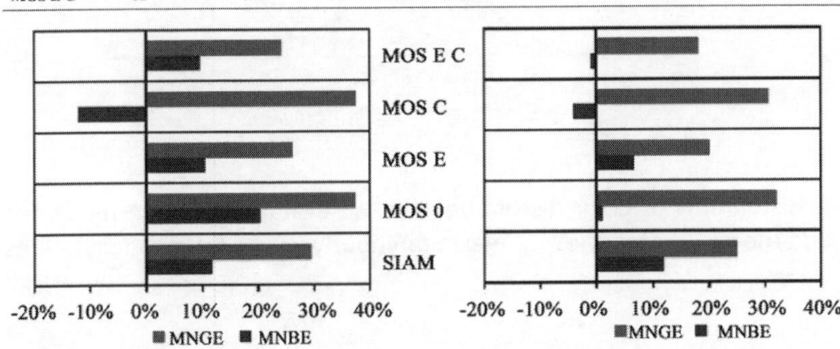

Figure 3. Comparison of MNGE and MNBE between the different MOS categories and the NO_2 modelling system for maximum 1-h value [left] and daily value [right].

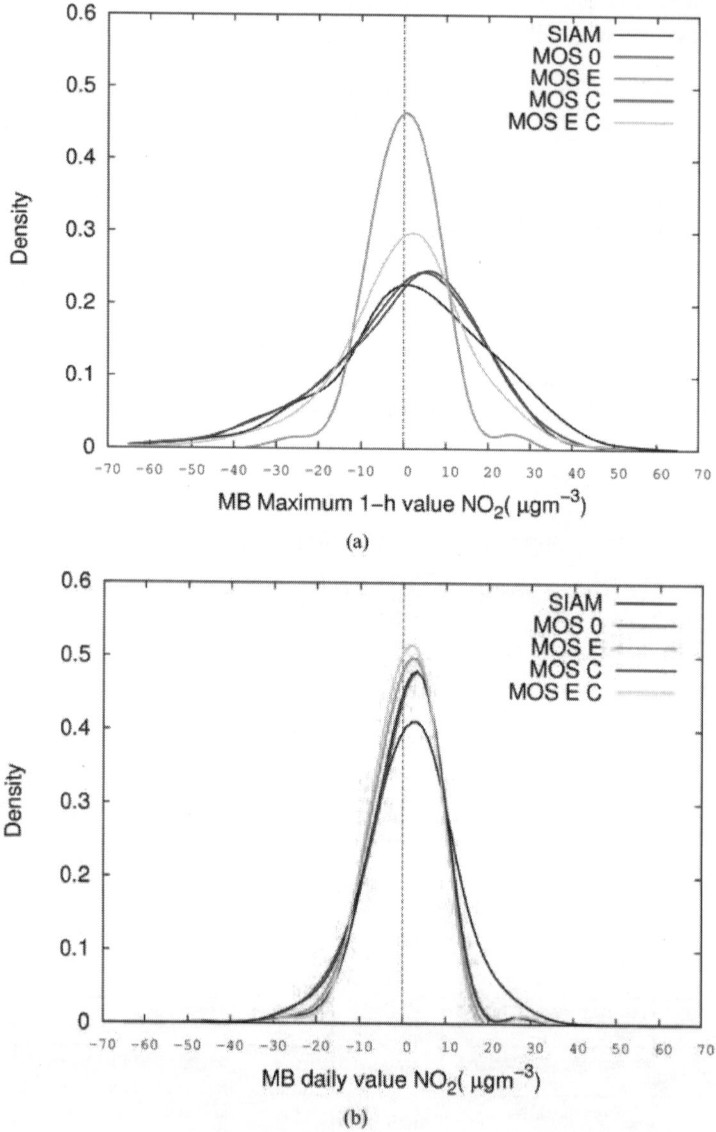

Figure 4. Histograms of error distributions in the evaluation of the modelling system and the MOS categories for maximum hourly values (a) and daily values (b) of NO_2.

As can be seen, the error distributions of both MOS E and MOS E C are the sharpest and narrowest and so the best. For daily values, the best results are attained by MOS E C but closely followed by the rest.

Through a Gaussian fit, we have calculated the width (twice the standard deviation) and displacement of the distributions with respect to the origin. The results are presented in Table 5. MOS E features a near zero shift and the narrowest width and MOS E C features the smallest displacement and width for daily values among all categories.

Figure 5(a) let us compare the time evolution of the mean daily values of observed maximum hourly values of NO_2 with those forecasted by the modelling system and those corrected by MOS E C. And Figure 5(b) shows the same comparison for daily values.

Paying attention at the behaviour of the errors on measurement stations we see that, in general, the improvement introduced by MOS E C is quite apparent in the maximum 1-h values, exhibiting generalized reductions in MNBE and MNGE and increases in IOA. MOS E C reduces MNBE in 78% of the measurement stations and 88% exceed [32] acceptance criteria for both MNBE and MNGE, which turned out to be only 69% and 75%, respectively, with the raw modelling system. On the other hand, the improvement produced by MOS E C over the modelling system in daily values is also very clear, reducing MNBE in 88% of measurement stations. Moreover, all stations exceed [32] acceptance criteria while 69% and 89% did, respectively, with the raw modelling system.

Table 5. Parameters of the Gaussian fit of the different MOS categories with respect to the maximum hourly value and daily value of NO_2.

Parameter	Maximum 1-h value		Daily value	
	Displacement (μ) ($\mu g \cdot m^{-3}$)	Width (2σ) ($\mu g \cdot m^{-3}$)	Displacement (μ) ($\mu g \cdot m^{-3}$)	Width (2σ) ($\mu g \cdot m^{-3}$)
SIAM	2.67	39.80	2.82	19.51
MOS 0	0.03	38.45	0.01	17.92
MOS E	0.12	17.20	0.06	15.09
MOS C	−0.12	37.90	−0.04	17.57
MOS E C	0.19	32.13	−0.01	13.79

3.2. PM_{10}

Table 6 shows the error analysis statistics of the different categories applied to the output of the modelling system of pollutant PM_{10}. Figure 6 displays the comparison between normalized mean absolute errors.

It can be seen that the best results are obtained with MOS E C. It decreases the underestimation produced by the modelling system, reducing MNBE from −24.2% to 5.9%, MNGE from 32.1 to 19.3% and increases IOA from 0.75 to 0.91.

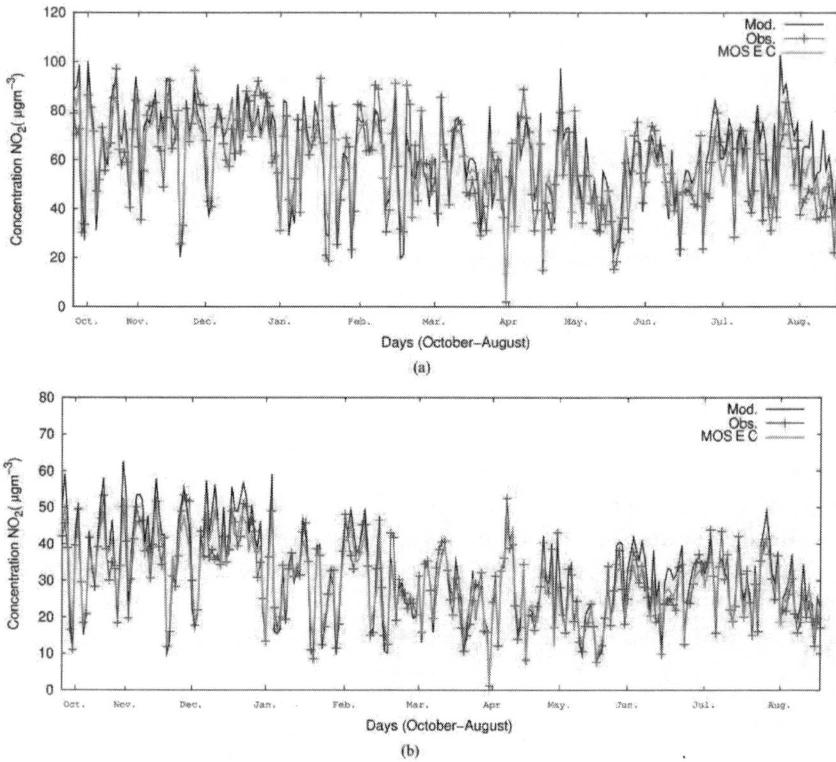

Figure 5. Time evolution of mean daily values of observed maximum hourly values of NO_2(a) and observed daily values (b) forecasted by the modelling system and corrected by MOS E C in the training period.

Figure 7 displays frequency histograms of the mean error distributions in forecasting daily values for different MOS categories and the modelling system. Table 7 shows the parameters of the Gaussian fit.

The results are very similar for categories MOS E and MOS E C for which the curves are narrower and centred.

To further show the improvement introduced by MOS E C, Figure 8 shows the comparison between the time evolution of daily PM_{10} values observed, forecasted by the raw modelling system and corrected by MOS E C.

Focusing on the impact of MOS E C at the measurement stations, it can be seen that MOS E C diminishes MNGE in all of them. Moreover, all measurement stations exceed [32] acceptance criteria for MNBE and MNGE, which is a very good achievement since only 21% and 58%, respectively, of the stations, satisfied the criteria with just the modelling system.

Table 6. Errors of all MOS categories as well as PM_{10} forecasting system.

	Daily value				
	Observed mean ($\mu g \cdot m^{-3}$)	Modelled mean ($\mu g \cdot m^{-3}$)	MNBE (%)	MNGE (%)	IOA
SIAM	17	13	−24.2	32.1	0.751
MOS 0	17	17	12.2	28.8	0.796
MOS E	17	17	6.9	21.7	0.884
MOS C	17	17	11.5	27.3	0.812
MOS E C	17	17	5.9	19.3	0.907

Table 7. Parameters of the Gaussian fit of the different MOS categories with respect to the daily value of PM_{10}.

Parameter	Daily value	
	Displacement (μ) ($\mu g \cdot m^{-3}$)	Width (2σ) ($\mu g \cdot m^{-3}$)
SIAM	4.46	10.87
MOS 0	0	10.56
MOS E	−0.07	6.03
MOS C	0.03	10.29
MOS E C	−0.01	7.97

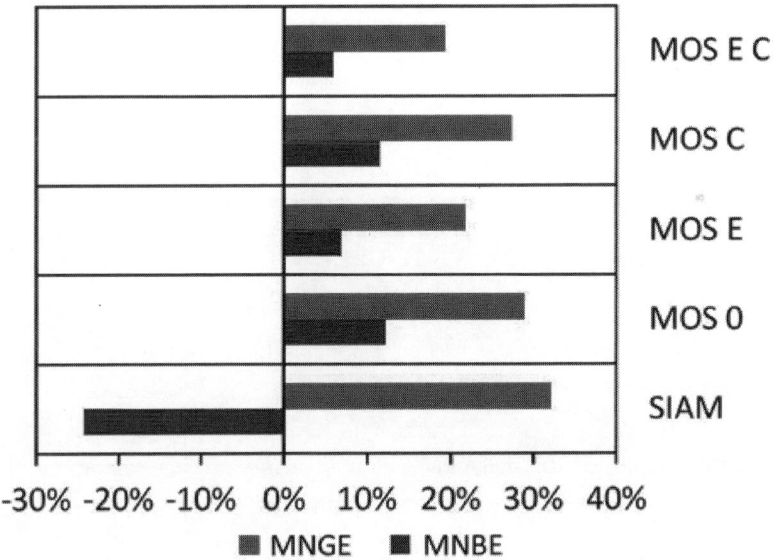

Figure 6. Comparison of MNGE and MNBE between the different MOS categories and the PM_{10} modelling system.

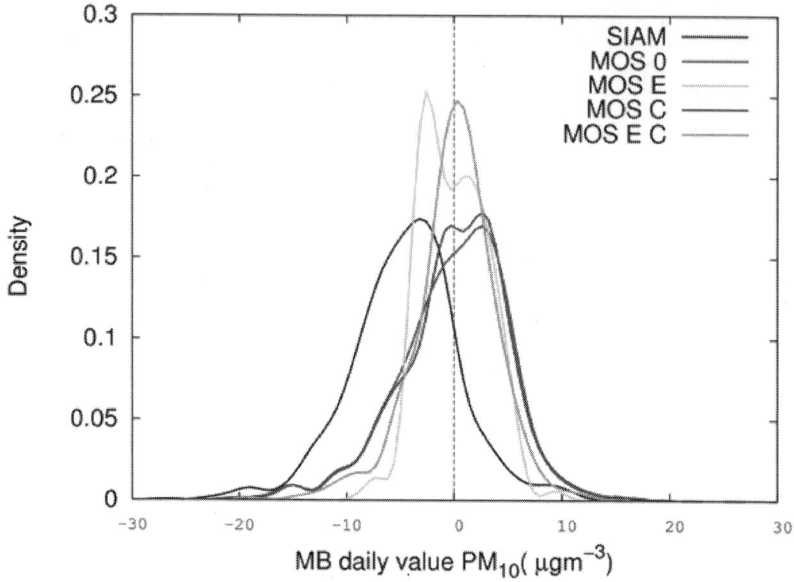

Figure 7. Histogram of error distributions in the evaluation of the modelling system and the MOS categories for daily values of PM_{10}.

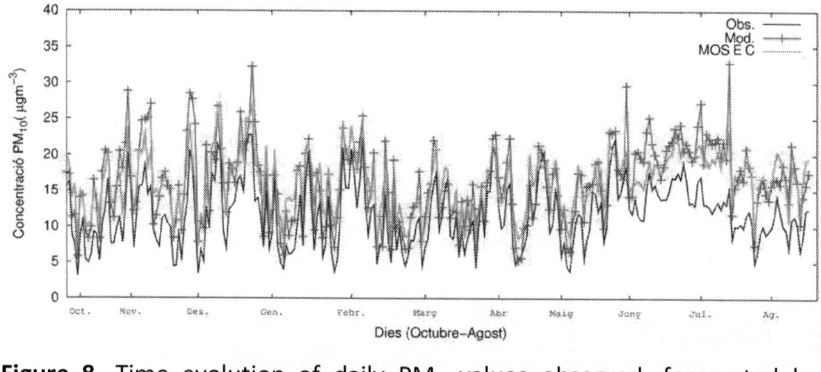

Figure 8. Time evolution of daily PM_{10} values observed, forecasted by the modelling system and corrected by MOS E C in the training period.

3.3. O_3

Table 8 shows the error analysis statistics of this pollutant for the different categories and Figure 9the comparison among the mean normalized errors.

As can be seen, MOS E C gets the best results, both in the maximum hourly value and maximum 8-hourly.

In the maximum hourly value, MOS E C slightly increases the MNBE of the modelling system from 1.8% to 2.8%, reduces MNGE from 20.8% to 13.2%, and increases IOA from 0.62 to 0.91.

For the maximum 8-h value, MOS E C reduces the MNBE of the modelling system from 7.0% to 3.9%, MNGE from 25.3% to 15.5%, and increases IOA from 0.68 to 0.92.

Figure 10 displays the mean error distributions of the forecasts of the values for every MOS category and the modelling system and Table 9 summarizes the errors of the Gaussian fit.

The best results are attained by MOS C and MOS E C, especially the former, for the maximum hourly value. MOS C produces the narrowest curve followed by MOS E C, though they are not quite centred. For the maximum 8-hourly value, the best results are attained by MOS E C followed by MOS C. MOS E C has the smallest width but is not the best centred.

The improvements achieved by MOS E C are further analyzed in Figure 11(a) and Figure 11(b) which compare the evolution of the averaged maximum 1-h and 8-h values of O_3 forecasted by the modelling system and corrected by MOS E C.

Table 8. Errors of all MOS categories as well as O_3 forecasting system.

	Maximum 1-h value					Maximum 8-h value				
	Observed mean ($\mu g \cdot m^{-3}$)	Modelled mean ($\mu g \cdot m^{-3}$)	MNBE (%)	MNGE (%)	IOA	Observed mean ($\mu g \cdot m^{-3}$)	Modelled mean ($\mu g \cdot m^{-3}$)	MNBE (%)	MNGE (%)	IOA
SIAM	91	86	1.8	20.8	0.618	81	79	7.0	25.3	0.678
MOS 0	91	91	7.6	22.0	0.623	81	81	9.7	25.8	0.702
MOS E	91	91	6.2	20.1	0.748	81	81	8.1	23.2	0.779
MOS C	91	90	4.4	16.6	0.845	81	81	6.3	20.1	0.851
MOS E C	91	91	2.8	13.2	0.910	81	81	3.9	15.5	0.917

Table 9. Parameters of the Gaussian fit of the different MOS categories with respect to the maximum 1-hourly value and maximum 8-hourly value of O_3.

Parameter	Maximum 1-h value		Maximum 8-h value	
	Displacement (μ) ($\mu g \cdot m^{-3}$)	Width (2σ) ($\mu g \cdot m^{-3}$)	Displacement (μ) ($\mu g \cdot m^{-3}$)	Width (2σ) ($\mu g \cdot m^{-3}$)
SIAM	−4.77	44.12	−2.25	42.45
MOS 0	−0.06	44.10	0.04	42.36
MOS E	0.17	17.56	0.14	38.82
MOS C	−0.16	34.87	−0.15	33.92
MOS E C	−0.24	28.50	−0.23	27.11

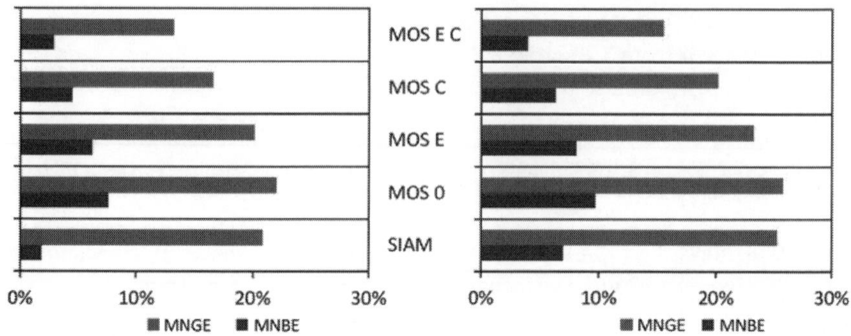

Figure 9. Comparison of MNGE and MNBE between the different MOS categories and the O_3 modelling system for maximum 1-h value [left] and maximum 8-h value [right].

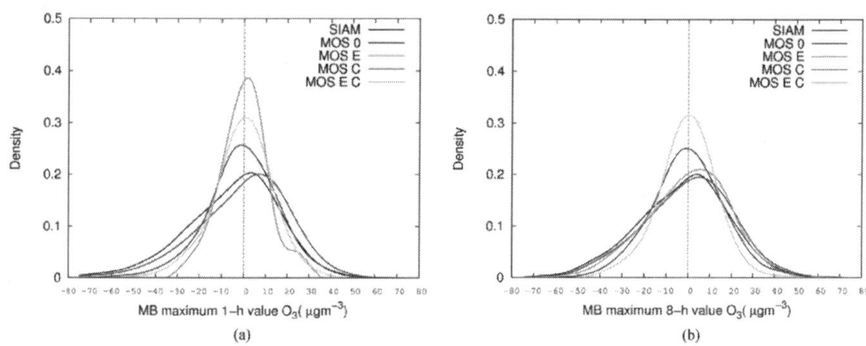

Figure 10. Histograms of error distributions in the evaluation of the modelling system and the MOS categories for maximum 1-hourly values (a) and maximum 8-hourly values (b) of O_3.

The analysis of errors by measurement stations show that the improvement introduced by MOS E C over the modelling system for maximum hourly values is apparent for most of the stations, with a generalized reduction of MNBE and MNGE, and an increase of IOA. MOS E C reduces MNGE in all stations and, moreover, all of them end up verifying [32] acceptance criteria for MNBE and MNGE.

Furthermore, the improvements produced by MOS E C over the modelling system, in case of the maximum 8- hourly value, are also generalized, with reductions in both MNBE and MNGE and increases in IOA. MOS E C reduces MNBE in all measurement stations and [32] criteria are met in all of them for MNBE and MNGE. This compares to 76% and 81%, respectively, of the measurement stations that met the criteria with the raw modelling system.

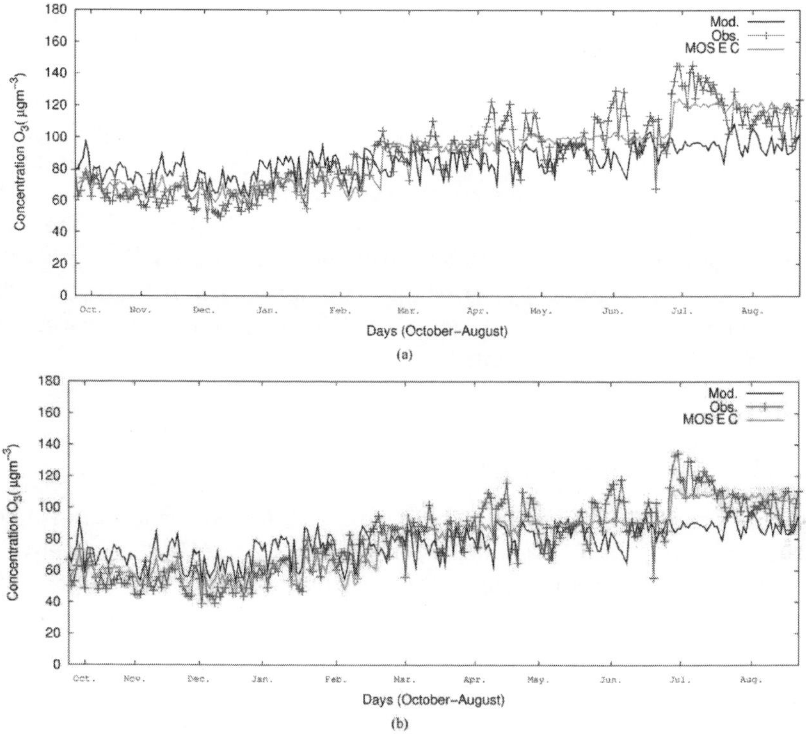

Figure 11. Time evolution of averaged maximum 1-hourly values (a) and maximum 8-hourly values (b) of O_3 observed, forecasted by the modelling system and corrected by MOS E C in the training period.

4. CONCLUSIONS

In this paper, we have applied the well-known MOS methodology in air quality modelling, a field where this technique is rarely employed. The output of an Eulerian coupled air quality modelling system feeds 4 MOS categories for each of these 5 statistics: maximum 1-h daily value and daily values of NO_2, daily values of PM_{10} and maximum 1-h and 8-h daily value of O_3; this amounts to 20 MOS considered.

The best results for all statistics have been obtained with the category that stratifies by measurement stations and by periods, as expected. The stratification by measurement stations corrects systematic errors of the modelling system. The stratification by periods allows for the treatment of the seasonal dependency of concentration values.

In summary, the application of the MOS methodology increases the accuracy of the maximum 1-h daily value of NO_2 from 71% to 75%, the daily value of

NO_2 from 75% to 82%, the daily value of PM_{10} from 68% to 81%, the 1-h daily value of O_3 from 79% to 87% and the 8-h daily value of $O_3$from 75% to 85%.

Our results highlight the improvements achieved by a quite simple mathematical tool with a very low computational cost. MOS methodology only increases a 2% the total computational cost of the operative air quality modeling system, and provides an improvement between 3% and 10% of the air quality forecasts. Nevertheless, one cannot forget that the MOS implementation should go in parallel with enhancements in the modelling system such as introducing better pollutants emission values to the modelling system. Any changes in the modelling system, however, imply the recalculation of the MOS regression coefficients.

ACKNOWLEDGEMENTS

This work was funded by the Catalan Government and the Spanish Government through the projects PTOP- 2013-608 and PTQ-12-05244 respectively. The authors gratefully acknowledge the technicians at the Territory and Sustainability Department of the Catalonian Government for providing air quality measurements.

REFERENCES

1. Newby, D.E., Manucci, P.M., Grethe, S.T., Baccarelli, A.A., Brook, R.D., Donaldson, K., Forastiere, F., Franchini, M., Franco, O.H., Graham, I., Hoek, G., Hoffmann, B., Hoylaerts, M.F., Künzli, N., Mills, N., Peklanen, J., Peters, A., Piepoli, M.F., Rajagopalan, S. and Storey, R.F. (2014) Expert Position Paper on Air Pollution and Cardiovascular Disease. European Heart Journal, in Press.

2. Crutzen, P.J. (2004) New Directions: The Growing Urban Heat and Pollution Island Effect-Impact on Chemistry and Climate. Atmospheric Environment, 38, 3539-3540.

3. Ritter, M., Müller, D., Tsai, M.-Y. and Parlow, E. (2013) Air Pollution Modeling over Very Complex Terrain: An Evaluation of WRF-Chem over Switzerland for Two 1-Year Periods. Atmospheric Research, 132-133, 209-222.

4. Querol, X., Viana, M., Moreno, T. and Alastuey, A. (2012) Bases científico-técnicas para un Plan Nacional de Mejora de la Calidad del Aire. Consejo Superior de Investigaciones Científicas (CSIC).

5. Soler, M.R., Hinojosa, J., Bravo, M., Pino, D. and Vila, J. (2004) Analyzing the Basic Features of Different Complex Terrain Flows by

Means of a Doppler Sodar and a Numerical Model: Some Implications to Air Pollution Problems. Meteorological Atmospheric Physics, 85, 141-154.

6. Aguirre-Basurko, E., Ibarra-Berastegui, I. and Madariaga, I. (2006) Regression and Multilayer Perceptron-Based Models to Forecast Hourly O3 and NO2 Levels in the Bilbao Area. Environmental Modelling Software, 21, 430-446.

7. McKeen, S., Wilczak, J. and Grell, G. (2005) Assessment of an Ensemble of Seven Real-Time Ozone Forecasts over Eastern North America during the Summer of 2004. Journal of Geophysical Research D, 110, Article ID: D21307.

8. Wilczak, J.M., McKeen, S. and Djalalova, I. (2006) Bias-Corrected Ensemble and Probabilistic Forecasts of Surface Ozone over Eastern North America during the Summer of 2004. Journal of Geophysical Research D, 111, Article ID: D23S28.

9. Delle Monache, L., Nipen, T., Deng, X., Zhou, Z. and Stull R. (2006) Ozone Ensemble Forecasts: 2. A Kalman Filter Predictor Bias Correction. Journal of Geophysical Research D, 111, Article ID: D05308.

10. Kang, D., Mathur, R., Rao, S.T. and Yu, S. (2008) Bias Adjustment Techniques for Improving Ozone Air Quality Forecasts. Journal of Geophysical Research, 113, Article ID: D23308.

11. Djalalova, J., Wilczak, J., McKeen, S., Grell, G., Peck-Ham, S., Pagowski, M., DelleMonache, L., McQueen, J., Tang, Y., Lee, P., McHenry, J., Gong, W., Bouchet, V. and Mathur, R. (2010) Ensemble and Bias-Correction Techniques for Air Quality Model Forecasts of Surface O3 and PM2.5 during the TEXAQS-II Experiment of 2006. Atmospheric Environment, 44, 455-467.

12. Arasa, R., Soler, M.R. and Olid, M. (2012) Numerical Experiments to Determine MM5/WRF-CMAQ Sensitivity to Various PBL and Land-Surface Schemes in North-Eastern Spain: Application to a Case Study in Summer 2009. International Journal of Environment and Pollution, 48, 105-116.

13. Dallavalle, J.P., Ericksonv, M.C. and Maloney, J.C. (2000) Model Output Statistics (Mos) Guidance for Short-Range Projections. National Weather Service/NOAA.

14. Nicolis, C. and Vannitsem, S. (2007) Dynamical Properties of Model Output Statistics Forecasts. Institute Royal Meteo-rologique de Belgique, Brussels.

15. Glahn, B., Peroutka, M. and Wiedenfeld, J. (2008) MOS Uncertainty Estimates in an Ensemble Framework. NOAAN, Wakeld, VA.

16. Monarski, A. (2011) Verification of Model Output Statistics Forecasts Associated with the North American Mesoscale Model Upgrade. Master's Scholarly Paper, Department of Atmospheric and Oceanic Science, University of Maryland, College Park.

17. Ranaboldo, M. (2011) Multiple Linear Regression MOS for Short-Term Wind Power Forecast. Universitat de Barcelona, Barcelona.

18. Karl, T.R. (1978) Potential Application of Model Output Statistics to Forecast of Surface Ozone Concentrations. National Meteorological Center's Limited Area.

19. Jianming, X. and Xiangde, X. (2010) Study of Statistically Correcting Model CMAQ-MOS for Forecasting Regional Air Quality. Nanjing University of Information Science Technology, Nanjing.

20. Pijnappel, A. (2011) Statistical Post Processing of Model Output from the Air Quality Model LOTOS-EUROS. De Bilt, Stageverslag.

21. Arasa, R., Picanyol, M. and Solé, J.M. (2013) Analysis of the Integrated Environmental and Meteorological Forecasting and Alert System (SIAM) for Air Quality Applications over Different Regions of the Iberian Peninsula. http://www.harmo.org/Conferences/Proceedings/_Madrid/downloadItem.asp?CID=H15-70

22. Denby, B. (2010) Guidance on the Use of Models for the European Air Quality Directive. A Working Document of the Forum for Air Quality Modelling in Europe FAIR-MODE, ETC/ACC Report.

23. Miles, J. and Shelvin, M. (2001) Applying Regression Correlation. Sage Publications, London.

24. Pérez López, C. (2003) Estadística. Problemas resueltos y problemas. Editorial Pearson.

25. Wilks, S.D. (2006) Statistical Methods in the Atmospheric Sciences. Second Edition, Academic Press, Waltham.

26. Montgomery, D.C., Peck, E.A. and Vining, G.G. (2011) Introducció al Análisis de Regressión Lineal. Grupo Editorial Patria. Tercera edición.

27. Chatterjee, S., Hadi, A.S. and Price, B. (2012) Regression Analysis by Examples. Fifth Edition, Wiley Series in Probability and Statistics.

28. Willmott, C.J., Ackleson, S.G., Davis, R.E., Feddema, J.J., Klink, K.M., Legates, D.R., O'Donnell, J. and Rowe, C.M. (1985) Statistics for the

Evaluation and Comparison of Models. Journal of Geophysical Research, 90, 8995-9005.

29. EPA (2009) Reassessment of the Interagency Workgroup on Air Quality Modeling (IWAQM) Phase 2 Summary Report: Revisions to Phase 2 Recommendations. Draft for Internal Review. US EPA Report No. EPA-454/R-09-XXX, Office of Air Quality Planning and Standards, Research Triangle Park, North Carolina.

30. Arasa, R. (2010) Avaluació del rendiment del sistema de modelització MM5/MNEQA/CMAQ de la qualitat de l'aire per a pronosticar l'ozó a Catalunya. Tethys núm. 7.

31. Arasa, R. (2011) Modelització i simulació fotoquíimica mesoescalar del transport del material particular i gasos a l'atmosfera. Tesis doctoral, Universitat de Barcelona, Barcelona.

32. EPA (1991) Guideline for Regulatory Application of the Urban Airshed Model. US EPA Report No. EPA-450/4-91-013. Office of Air and Radiation, Office of Air Quality Planning and Standards, Research Triangle Park, North Carolina.

33. EPA (2005) Guidance on the Use of Models and Other Analyses in Attainment Demonstrations for the 8-Hour Ozone NAAQS. US EPA Report No. EPA-454/R-05-002, Office of Air Quality Planning and Standards, Research Triangle Park, North Carolina.

Chapter 8

Water Quality Assessment of a Tropical Mexican Lake Using Multivariate Statistical Techniques

Jessica Badillo-Camacho[1], Eire Reynaga-Delgado[2], Isela Barcelo-Quintal[3], Pedro F. Zarate del Valle[1], Ulrico J. López-Chuken[4], Eulogio Orozco-Guareño[1], Jorge Israel AlvarezBobadilla[3], Sergio Gomez-Salazar[5*]

[1]*Departamento de Química, Universidad de Guadalajara-Centro Universitario de Ciencias Exactas e Ingenierías, Guadalajara, México*

[2]*Departamento de Farmacobiologia, Universidad de Guadalajara-Centro Universitario de Ciencias Exactas e Ingenierías, Guadalajara, México*

[3]*Departamento de Ciencias Básicas, Universidad Autónoma Metropolitana, Unidad Azcapotzalco, México D.F., México*

[4]*Facultad de Ciencias Químicas, Universidad Autónoma de Nuevo León, San Nicolás de los Garza, Mexico*

[5]*Departamento de Ingeniería Química, Universidad de Guadalajara-Centro Universitario de Ciencias Exactas e Ingenierías, Guadalajara, México*

ABSTRACT

Water quality of Mexican tropical lake Chapala was assessed through multivariate statistical techniques, cluster analysis (CA) and principal component analysis (PCA) at ten different monitoring sites for ten physicochemical variables and six metals. This study evaluated and interpreted complex water quality data sets and apportioned of pollution sources to get better information about water quality. From descriptive statistics results, the highest concentrations of metals occurred during the dry season, and this trend was explained by the fact that an unusual rainy event occurred during the month of February 2009 and brought metals into the lake by runoffs from nearby mountains. According to international criteria for water consumption by aquatic organisms [USEPA], only Zn concentration values were below these criteria whereas the values of Ni, Pb, Cd and Fe were above the corresponding values set in these criteria (Ni: 52 μg·L^{-1}, Pb: 2.5 μg·L^{-1}, Cd: 0.25 μg·L^{-1}, and Fe: 1000 μg·L^{-1}). The correlations were observed by PCA, which were used to classify the samples by CA, based on the PCA scores. Seven significant cluster groups of sampling locations—(sites 4 and 5), (sites 3 and 9), (site 7), (site 10), (sites 2 and 6), (site 8) and (site 1)— were detected on the basis of similarity of their water quality. The results revealed that the stress exerted on the lake caused by waste sources follows the order: domestic > agricultural > industrial.

Keywords: Multivariate Statistical Analysis, toxic Metals, Water Quality, Lake Pollution

1. INTRODUCTION

The pollution of water sources such as lakes and rivers is a problem of global concern in large part because many of these water bodies are used as water supply sources for human consumption [1] . This is because the uncontrolled human settlements have an impact on the water quality of these water bodies due to the lack of infrastructure (e.g. water treatment plants) that permits the appropriate treatment of domestic, municipal or industrial wastes (especially in the underdeveloped countries). Within this context, a main indicator controlling health and state of disease in humans and animals is the water quality. In general, natural processes and anthropogenic inputs determine water quality of these surface water bodies [2] - [4] . In the former, runoffs inputs are an example of these processes and come from seasonal events that are affected by the climate (weathering and soil erosion) [5] ; the latter ones include discharges from municipal, industrial and agricultural activities and constitute a permanent source of pollutants, especially if the discharges are not treated appropriately [6] [7] .

Several research works have been focused on the anthropogenic contamination of ecosystems [8] [9] . However, monitoring programs that provide representative and reliable information of the data are not easily implemented due to both spatial and temporal variations in water quality [10] . This is due in part to the large amount of data generated from the measured physicochemical variables, which in turn makes difficult to interpret and to draw meaningful conclusions. The use multivariate statistical techniques, such as principal component analysis (PCA), factor analysis (FA), and cluster analysis (CA) provide a valuable tool in the interpretation of complex data matrices of the water quality and the ecological status of a region (or water body) of interest. Consequently, the use of these techniques can help in the reliable management of water resources and provide a fast solution when facing pollution problems, especially those of water bodies used for human consumption [11] [12] .

Shallow tropical Lake Chapala is Mexico's largest freshwater body, and is the most important lake in this country, in part because itis used a water supply resource by the city of Guadalajara. Pollution of this water body has prompted concern since large quantities of wastes from municipal, industrial, and agricultural activities from the entire Lerma-Chapala basin flow largely untreated into the lake through the Lerma River (the main tributary to the lake, see Figure 1) [13] [14] . Therefore, the main objective of the present study was to assess the water quality of this lake through concentrations of the macronutrients (nitrate-N, nitrite-N, total nitrogen, and total phosphorus), other physicochemical variables (pH, alkalinity, total suspended solids, conductivity and chemical oxygen demand), and heavy metals (Fe, Cd, Ni, Zn, Cr, and Pb). Multivariate techniques such as PCA, CA and FA were applied to the large data sets obtained during a two-period sampling program (one during the dry season and other at the end of the rainy season of 2009) to find similarities and dissimilarities among the different sampling sites, and to ascertain the influence of the pollution sources on the water quality variables.

2. MATERIALS AND METHODS

2.1. Reagents and Instruments

Analytical grade reagents were used without further treatment. Metal concentrations were analyzed by Atomic Absorption Spectrometry (AAS, Analytik Jena Model ContrAA 300, Jena, Germany). Certified standard metal ion solutions (Sigma-Aldrich, St. Louis, MO.) were used to obtain concentration versus absorbance calibration curves. Instrument detection limits were ($mg \cdot L^{-1}$): 1) Cd = 0.002, 2) Pb = 0.05, 3) Cr = 0.02, 4) Mn=0.01, 5) Zn = 0.005, 6) Ni =0.02, 7) Cu = 0.01, 8) Fe = 0.02. Blank solutions, i.e., the same composition as the target solution but without metal ions, were used as background for absorption corrections. All the solutions were prepared using deionized water with a resistivity of 18 MΩ (Barnstead, Chicago, Ill). A Thermo Electron (Model Orion 3- Star) pH/conductivity meter was used for pH and conductivity measurements.

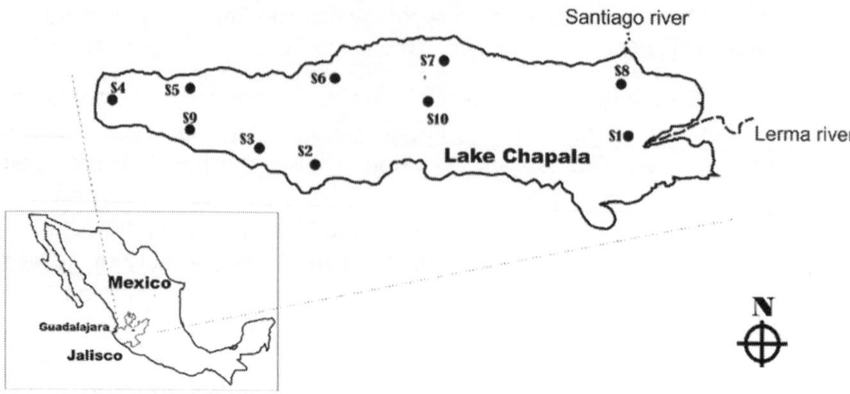

Figure 1. Location of water sampling sites along Lake Chapala. S1: Lerma; S2: Tuxcueca; S3: Soyatlán; S4: Jocotepec; S5: San Juan Cosalá; S6: Aqueduct; S7: Mezcala; S8: Jamay; S9, Middle; S10, Depocenter.

2.2. Study Zone

Lake Chapala is part of the Lerma-Chapala watershed and is located in the western central side of Mexico at about 1515 m above sea level; it covers a surface area of ~80 × 15 km^2, and its average depth varies between 4 and 7 m (Figure 1). The lake provides ~7.5 m^3/s of the water consumed by the city of Guadalajara (pop. ~4 million, see Figure 1) [15] , and has a unique ecosystem. Due to wind action and low depth, lake stratification is reduced, consequently the lake is oxidized at all depths [16] , as reflected by the dissolved oxygen values (6.8 - 7.0 $mg \cdot L^{-1}$) which represent a saturation level > 90%. This phenomenon is attributed to the almost complete water mix and to the low depth of the lake [16] [17] . The human activities have changed significantly the original regime of the lake over the last 50 years. These changes have been caused by inputs of the Lerma River due to the industrial development of cities located along this river. The lake acts as a sink for dissolved substances (e.g. heavy metals) and suspended sediments transported from up-stream agricultural, urban and industrial areas through the Lerma River and the local watershed [18] .

In this study, all sampling sites were selected to cover a wide range of variables of key sites, which reasonably represent the water quality of the lake, accounting for the tributary and inputs from wastewater drains that have impact on the water quality. Ten sites were selected during the dry season of 2009 (May) and at the end the rainy season of 2009 (November) to collect water samples (Figure 1). The sites were selected on the basis of their potential of pollution due to anthropogenic activities that took place at nearby towns and they were distributed at intervals around this lake for representativeness and to account for major pollution sources. Sites were identified depending on when the sampling campaign was conducted, for example: S1A corresponded to Site 1 sampled during the dry season (May, 2009) and S1B represented Site 1 sampled after rainy season (November, 2009); the rest of sites were identified using the same notation.

2.3. Water samples collection and analysis

Water samples were manually collected from approximately 30 cm beneath the water surface in polyethylene bottles previously soaked in 10% v/v HNO_3 for 24 h and rinsed several times with deionized water. All samples were cooled at 4°C and transported to the laboratory for further analysis. Temperature and pH were measured in the field. The analyses of physicochemical variables (NO_2^-, NO_3^-, PO_4^{3-}, etc.) for water quality followed protocols established by the Mexican Standard [19] and APHA [20] . For metal analysis, samples were acidified in the field with concentrated HNO_3 to pH < 2. Samples were digested according to the APHA Method 3030 K: Microwave-Assisted Digestion [20] in a programmable Microwave Oven Model Ethos Touch (Sorisole, Italy) and the extracts analyzed for heavy metal concentrations Atomic Absorption Spectrometry. Analytical blanks were analyzed in a way similar to that of the samples and metal concentrations were determined using the standard solutions prepared in the same acid matrix.

2.4. Statistical Methods

Descriptive statistics (mean, standard deviation and maximum) were calculated for the physicochemical variables in water samples. The Pearson correlation coefficient was obtained to describe the degree of association between two variables under study. FA was employed to establish the possible relationships between the physicochemical variables of the sites sampled. This multivariate method was used here to obtain information about: 1) the most relevant characteristics of the physicochemical variables with a minimal loss of original data [1] [21] [22] , 2) to create an entirely new set of factors much smaller in number when compared to the original data set of variables focused on reducing the contribution of the less significant variables to simplify even more the data structure coming from the principal component analysis [23] . CA was used to explore the similarities between water samples [24] and grouping the sites according to the similarity of contaminants. Hierarchical agglomerative CA was performed on the normalized data set using squared Euclidean distances as a measure of similarity. The CA technique is a classification procedure that involves measuring either the distance or the similarity between the objects to be clustered. Details of this method can be found elsewhere [25] [26] . All statistical tests were made using StatGraphics Centurion XV software (StatPoint Inc., USA).

2.5. Quality Assurance

All glassware and glass fiber filters were rinsed using deionized water with a resistivity of 18 MΩ (Barnstead, Chicago, Ill) and heated at 450°C prior to use. The Relative Standard Deviation (RSD), as determined by running a standard 10 times, calculating the average, and dividing it into the standard deviation, was less than 4% across all analyses.

3. RESULTS AND DISCUSSION

3.1. Water Analysis Results

The results of the measured water quality variables are shown in Table 1 for the two sampling seasons. The pH values are alkaline (8.6 - 9.2); site S8A presents the highest value. This site is characterized as a place that receives the Lerma River inputs directly. The mean pH value was 8.8. This value lies within the appropriate range for aquatic life survival [27] . Conductivity values were higher on sites sampled after the rainy season compared to the dry season campaign; sites S1B and S8B presented the highest values (840 and 860 $\mu S \cdot cm^{-1}$, respectively). Alkalinity values were highest at sites S2A and S7A (318 and 340 $mg \cdot L^{-1}$, respectively). These values can be attributed to the fact that bicarbonate ion entering the lake is dissociated to carbon dioxide which in turn is consumed by algae for their synthesis needs [28] . Sites S2A and S7A are located in a zone where tourism activities take place and there are wastewater discharges mainly from domestic origin that provide the appropriate media for algae proliferation. Nutrient variables such as NO_2^- and PO_4^{3-} show significant variations from one season to the other. For example, NO_2^- values could be detected at sites S1A, S4A and S6A; NO_3^- values ranged between 1.7 and 12.5 $mg \cdot L^{-1}$ with the highest values found after the rainy season. PO_4^{3-} concentrations were in the range of 0.3 - 0.7 $mg \cdot L^{-1}$, with highest values measured during the dry season. A significant decrease of N_T, NO_2^- and NO_3^- was observed after the rainy season with undetected values at site S6B. The highest N_T values were measured at site S4A during the dry period and are attributed to the fact that this site is located in a hilly zone where there is an intense raspberry and blueberry agricultural activity and the use of large amounts of N- and P-rich pesticides is generalized that end up in the lake

Table 1. Descriptive statistics of ten variables and six metals (as total metal) measured in waters of ten sites of Lake Chapala. Concentrations in mg·L^{-1}; conductivity S in µS·cm^{-1}.

Site	pH	S	NO_2^-	NO_3^-	N_T	TS	PO_4^{3-}	COD	Alk*	SO_4^{2-}	Fe	Cd	Ni	Zn	Cr	Pb
S1A	8.9	528.0	0.1	1.9	12.7	651.0	0.7	11.0	290.0	79.3	ND	0.2	ND	ND	0.6	1.3
S1B	8.6	840.0	ND	8.7	12.0	644.0	0.7	23.0	262.6	64.0	1.7	ND	ND	ND	ND	ND
S2A	9.0	126.0	ND	1.7	1.8	591.0	0.7	24.0	318.0	69.8	0.5	ND	ND	ND	ND	1.3
S2B	8.6	740.0	ND	8.3	4.0	608.0	0.5	17.0	282.7	80.0	0.5	ND	ND	ND	ND	ND
S3A	9.0	528.0	ND	1.7	1.7	562.0	0.6	14.0	282.0	79.3	2.1	ND	ND	ND	ND	ND
S3B	8.7	760.0	ND	12.5	1.0	592.0	0.4	22.0	291.5	82.0	1.0	ND	ND	ND	ND	ND
S4A	8.9	524.0	0.1	1.9	15.6	586.0	0.7	15.0	280.0	72.0	2.9	0.2	1.0	ND	ND	1.1
S4B	8.7	770.0	ND	7.1	8.0	600.0	0.5	34.0	292.1	79.0	0.7	ND	ND	ND	ND	ND
S5A	9.0	522.0	ND	1.8	3.5	573.0	0.7	20.0	284.0	76.5	2.2	0.2	1.0	ND	0.6	1.3
S5B	8.7	760.0	ND	8.8	15.0	576.0	0.4	22.0	286.8	79.0	0.7	ND	ND	ND	ND	ND
S6A	9.0	174.0	0.1	1.8	3.6	583.0	0.7	20.0	286.0	71.3	0.8	ND	ND	0.1	ND	ND
S6B	8.6	750.0	ND	ND	ND	608.0	0.4	20.0	276.7	80.0	1.5	ND	ND	ND	ND	ND
S7A	8.7	227.0	ND	1.7	4.5	584.5	0.5	9.0	340.0	77.9	0.8	0.2	ND	ND	ND	ND
S7B	8.7	670.0	ND	8.6	2.0	552.0	0.4	19.0	170.9	79.0	1.0	ND	ND	ND	ND	ND
S8A	9.2	527.0	ND	4.6	1.3	581.0	0.6	8.0	284.0	76.3	2.8	ND	ND	ND	ND	1.3
S8B	8.8	860.0	ND	8.1	1.0	632.0	0.6	22.0	278.7	73.0	2.1	ND	ND	ND	ND	ND
S9A	9.0	524.0	ND	1.8	15.0	569.0	0.5	4.0	246.0	78.9	2.0	ND	ND	ND	ND	ND
S9B	8.7	760.0	ND	6.3	3.9	576.0	0.3	16.0	279.7	80.0	1.2	ND	ND	ND	ND	ND
S10A	9.0	389.0	ND	1.7	1.9	563.0	0.7	6.0	280.0	75.8	1.8	0.2	1.0	0.1	ND	1.3
S10B	8.6	740.0	ND	8.4	10.0	576.0	0.4	18.0	266.7	80.0	1.2	ND	ND	ND	ND	ND
Mean	**8.8**	**586.0**	**ND**	**4.9**	**5.9**	**590.4**	**0.6**	**17.2**	**278.9**	**76.7**	**1.4**	**ND**	**0.2**	**ND**	**0.1**	**0.4**
Min	**8.6**	**126.0**	**ND**	**ND**	**ND**	**552.0**	**0.3**	**4.0**	**170.9**	**64.0**	**ND**	**ND**	**ND**	**ND**	**ND**	**ND**
Max	**9.2**	**860.0**	**0.1**	**12.5**	**15.6**	**651.0**	**0.7**	**34.0**	**340.0**	**82.0**	**2.9**	**0.2**	**1.0**	**0.1**	**0.6**	**1.3**

*Alkalinity as HCO_3^-. ND = Not detected. Site key: A = sampled during dry season (May of 2009); B = sampled at the end of rainy season (Nov. 2009).

possibly through illegal discharges. In the case of the total solids, an approximately constant trend was observed for all sites during the two sampling seasons, except for site S1A. This site is located at the entrance of the Lerma River which in turn is the main contributor of suspended particles into Lake Chapala. COD values were higher after the rainy season; site S4B presented the highest value (34 mg·L^{-1}) whereas site S9A presented the lowest value (4 mg·L^{-1}) and the mean value was 17.20 mg·L^{-1} for the two sampling campaigns. SO_4^{2-} values showed little variation throughout the lake (64 - 82 mg·L^{-1}) during the two campaigns, with a mean value of 76.7 mg·L^{-1} and the highest concentrations obtained after the rainy season campaign. In the case of metals, Fe and Pb were the two elements prevailing during the two seasons;

the maximum Fe value was measured at site S4A (2.9 mg·L^{-1}) and Pb was present in the waters at sites S1A, S2A, S5A, S8A and S10A (1.3 mg·L^{-1} at each site); Cr was detected at sites S1A and S5A. Cd, one of the most toxic metals, presented concentrations of 0.2 mg·L^{-1} at sites S1A, S4A, S7A and S10A. In general, the highest concentrations of metals occurred during the dry season, and this trend can be explained by the fact that an unusual rainy event occurred during the month of February 2009 and brought metals into the lake by runoffs from nearby mountains. The increase of metal concentrations in surficial water bodies during unusual rainy events has been reported by others (Roberto et al. 2008; Gupta et al. 2010). According to international criteria for water consumption by aquatic organisms [29] , only Zn concentration values were below these criteria whereas the values of Ni, Pb, Cd and Fe were above the corresponding values set in these criteria (Ni: 52 µg·L^{-1}, Pb: 2.5 µg·L^{-1}, Cd: 0.25 µg·L^{-1}, and Fe: 1000 µg·L^{-1}).

Table 2. Correlation analysis of ten variables and six metals in ten sites of Lake Chapala.

	pH	S*	NO_2^-	NO_3^-	NT	TS	PO_4^{3-}	COD	Alk	SO_4^{2-}	Fe	Cd	Ni	Zn	Cr	Pb
pH	1.00															
S	−0.61*	1.00														
NO_2^-	0.61*	−0.51	1.00													
NO_3^-	−0.57	0.69*	−0.55	1.00												
N_T	−0.07	0.15	0.25	0.03	1.00											
TS	−0.26	0.30	0.11	0.13	0.18	1.00										
PO_4^{3-}	0.61*	−0.50	0.70*	−0.48	0.09	0.28	1.00									
COD	−0.41	0.36	−0.26	0.46	−0.09	0.27	−0.16	1.00								
Alk	0.14	−0.39	0.20	−0.29	−0.07	0.26	0.20	0.02	1.00							
SO_4^{2-}	−0.25	0.26	−0.43	0.14	−0.13	−0.35	−0.67*	−0.18	−0.09	1.00						
Fe	0.43	0.09	0.31	−0.21	−0.03	−0.23	0.17	−0.32	−0.16	−0.25	1.00					
Cd	0.23	−0.41	0.63*	−0.50	0.18	0.02	0.47	−0.42	0.31	−0.04	0.11	1.00				
Ni	0.35	−0.21	0.55	−0.36	0.08	−0.27	0.46	−0.22	0.03	−0.18	0.49	0.72*	1.00			
Zn	0.17	−0.32	0.31	−0.12	−0.13	−0.19	0.35	−0.10	−0.16	−0.33	−0.05	0.08	0.26	1.00		
Cr	0.24	−0.11	0.42	−0.29	0.12	0.26	0.39	−0.10	0.09	0.09	−0.10	0.59	0.36	−0.14	1.00	
Pb	0.66*	−0.47	0.58	−0.48	0.00	0.01	0.63*	−0.29	0.23	−0.25	0.24	0.60*	0.62*	0.03	0.54	1.00

*$p < 0.01$; *S = conductivity.

3.2. Statistical Analysis Results

The correlations found between the monitored variables were few, and in general, positive (seeTable 2). The greatest number of correlations occurred for pH with a negative value between this variable and conductivity and slightly positive with NO_2^- and Pb. Other significant positive correlations were found between Pb and Cd (0.6), Pb and Ni (0.62). Finally, the conductivity presented weakly negative correlations with metals, except with Fe.

3.3. Factor Analysis Results

After varimax rotation and Principal Component Analysis, six factors were extracted that represented 81.8% of variability of original data. Weights for each of the variables for each factor are shown inTable 3 and Figure 2 (bi-plots). Factor F1 (34.7% of variance) was comprised of variables associated to domestic-type wastewaters and agricultural runoffs (pH, Conductivity S, COD, NO_3^-); factor F2 (12.9% of variance) was formed by heavy metals (Cd, Cr, Pb, and Ni); factor F3 (11% of variance) included variables characteristic of agricultural runoffs (SO_4^{2-} and PO_4^{3-}) and suspended particles (TS) coming from erosion, human and agricultural activities; factors F4 (9.8% of variance) and F5 (6.9% of variance) were comprised of Fe and Zn, respectively; factor F6 (6.3% of variance), included T_N which is a nutrient associated with surficial water eutrophization. Based on the results of factor analysis, it can be inferred that the contaminants that exert the highest environmental stress on Lake Chapala, are the wastes from domestic origin, followed by heavy metals from either natural source or from non- point discharges containing industrial residues. Finally, the contaminants that exert the least environmental stress are those from agricultural activities that take place along the lake and inputs from Lerma River.

3.4. Cluster Analysis Results

Methods of Ward and the Closest Neighbor were applied during the Cluster Analysis using averaged concentration values of each variable; the results are presented in Figure 3. The first cluster group contained sites S4 (Jo cotepec) and S5 (San Juan Cosala). These sites are located in zones with the highest agricultural and tourism activities, and consequently this cluster represents the sites with the highest environmental stress of Lake Chapala. The second cluster contains sites S3 and S9. These sites are located close to land places where agricultural activities are predominant. Third cluster is comprised of site S7, a site where touristic activity is lower than at sites S4 and S5. Fourth cluster contains site S10. This site is far from human settlements and from agricultural or industrial activities. Sites S2 and S6 comprise the fifth cluster. These sites are characterized by little commercial, agricultural or touristic activities. It is worth noting that site S6 is the place where the water intake is located that supplies water to the nearby city of Guadalajara. Clusters seven and eight comprise sites S8 and S1, respectively. These are the least contaminated sites of the lake. Based upon the results of the cluster analysis, it can be established that the source of environmental stress on Lake Chapala arises mainly from the activities that take place at populations located along the lakeshore. In this regard, it can be seen that the stress exerted on the lake caused by waste sources follows the order: domestic > agricultural > industrial.

Table 3. Factor loads after varimax rotation of ten variables and six metals at ten sites of Lake Chapala.

	F1	F2	F3	F4	F5	F6
pH	0.69	0.21	0.23	0.44	−0.03	−0.13
S**	−0.89	−0.13	−0.13	0.16	0.23	0.15
NO_2^-	0.46	0.52	0.45	0.17	−0.16	0.23
NO_3^-	−0.78	−0.30	−0.06	−0.09	−0.02	−0.02
NT	−0.07	0.09	0.12	−0.01	0.06	0.91
TS	−0.29	0.10	0.60	−0.35	0.45	0.19
PO_4^{3-}	0.40	0.44	0.70	0.09	−0.15	0.03
COD	−0.60	−0.14	0.35	−0.30	0.07	−0.34
Alk	0.46	0.11	0.22	−0.37	0.47	−0.17
SO_4^{2-}	0.10	0.05	0.91	0.16	0.18	0.05
Fe	0.09	0.07	0.11	0.94	0.05	0.01
Cd	0.34	0.83	−0.02	−0.04	−0.04	0.20
Ni	0.12	0.74	0.06	0.42	−0.30	−0.02
Zn	0.16	0.02	0.25	−0.10	−0.89	−0.07
Cr	0.02	0.84	0.02	−0.15	0.20	0.05
Pb	0.41	0.69	0.23	0.23	0.09	−0.14

**S = conductivity.

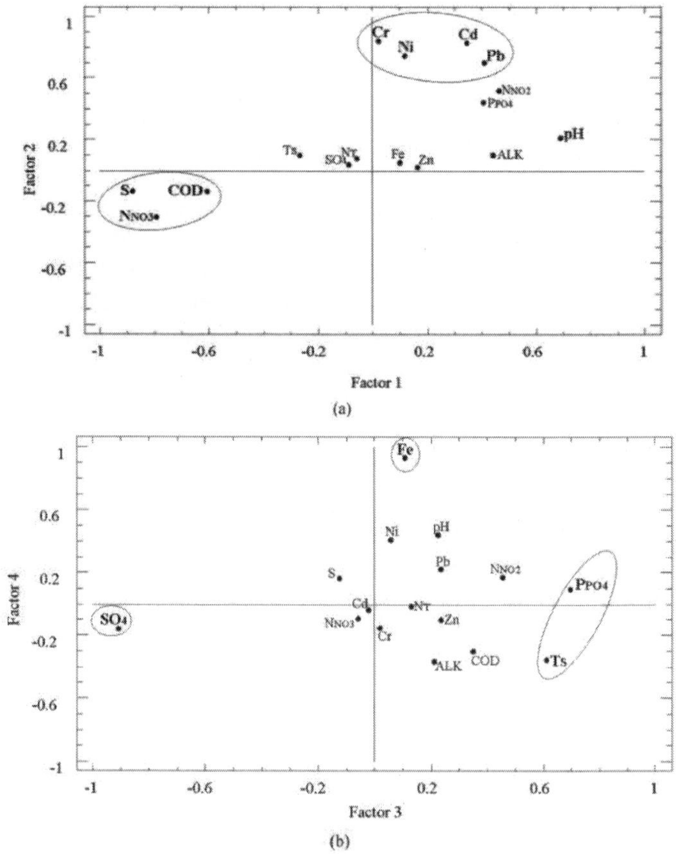

Figure 2. Factor analysis bi-plot of variables measured in water samples of Lake Chapala. (a) Factor 1 vs. Factor 2. (b) Factor 3 vs. Factor 4. S = conductivity.

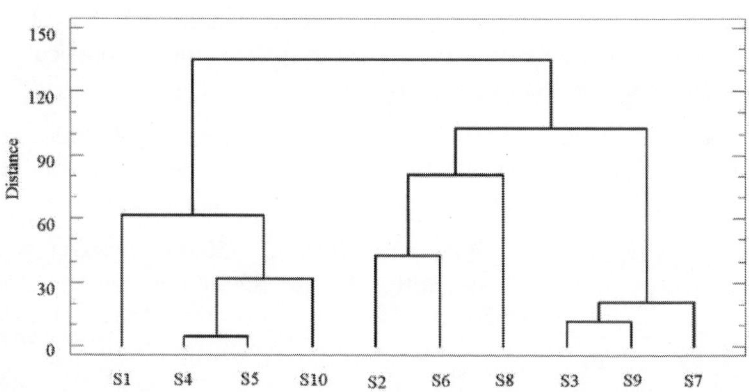

Figure 3. Cluster analysis results on water samples of ten sites of Lake Chapala.

4. CONCLUSION

The analyzed physicochemical variables of waters from Lake Chapala are within international values (WHO and USEPA). The presence of Cd, Ni and Pb (even in small amounts) in lake waters suggests the need of careful monitoring of the water supplied to the city of Guadalajara used for human consumption. Variables such as phosphates and nitrates exceed the corresponding limits set by WHO and USEPA. High concentrations of these nutrients were attributed to the widespread use of phosphate-rich fertilizers and detergents on parts of land along the lake shoreline and in some parts of the Lerma River close to the lake. A study is in progress to determine more accurately (via transport models) metal pollution point sources that would help to propose pollution control policies with the final objective of reducing the presence of toxic metals in the lake.

ACKNOWLEDGEMENTS

Jessica Badillo-Camacho, Eire Reynaga-Delgado, Isela Barcelo-Quintal, Pedro F. Zarate del Valle, Ulrico J. López-Chuken, Eulogio Orozco-Guareño, Jorge Israel AlvarezBobadilla, Sergio Gomez-Salazar (CONACyT) and the Ministry of Public Edu- cation-PROMEP for their support through grants No. 84252 and 103.5/13/9346, respectively, and for the scholarship of Jessica Badillo-Camacho from CONACyT. We acknowledge Ms. Mary Medina and Rebeca Sanchez (Universidad de Guadalajara) for their assistance in the measurement of water quality variables.

REFERENCES

1. Kazi, T.G., Arain, M.B., Jamali, M.K., Afridi, H.I., Sarfraz, R.A., Baig, J.A. and Shah, A.Q. (2009) Assessment of Water Quality of Polluted Lake Using Multivariate Statistical Techniques: A Case Study. Ecotoxicology and Environmental Safety, 72, 301-309.
2. Carpenter, S.R., Caraco, N.F., Correll, D.L., Howarth, R.W., Sharpley, A.N. and Smith, V.H. (1998) Nonpoint Pollu- tion of Surface Waters with Phosphorus and Nitrogen. Ecological Applications, 8, 559-568.
3. Jarvie, H.P., Whitton, B.A. and Neal, C. (1998) Nitrogen and Phosphorus in East Coast British Rivers: Speciation, Sources and Biological Significance. Science of the Total Environment, 210-211, 79-109.

4. Ravichandran, S. (2003) Hydrological Influences on the Water Quality Trends in Tamiraparani Basin, South India. En- vironment Monitoring and Assessment, 87, 293-309

5. Karlsson, K., Viklander, M., Scholes, L. and Revitt, M. (2010) Heavy Metal Concentrations and Toxicity in Water and Sediment from Storm Water Ponds and Sedimentation Tanks. Journal of Hazardous Materials, 178, 612-618.

6. Singh, K.P., Malik, A., Mohan, D. and Sinha, S. (2004) Multivariate Statistical Techniques for the Evaluation of Spa- tial and Temporal Variations in Water Quality of Gomti River (India)—A Case Study. WaterResearch, 38, 3980-3992.

7. Vega, M., Pardo, R., Barrado, E. and Deban, L. (1998) Assessment of Seasonal and Polluting Effects on the Quality of River Water by Exploratory Data Analysis. Water Research, 32, 3581-3592.

8. Szymanowska, A., Samecka-Cymerman A. and Kempers, A.J. (1999) Heavy Metals in Three Lakes in West Poland. Ecotoxicology and Environmental Safety, 43, 21-29.

9. Issa, Y.M., Elewa, A.A., Rizk, M.S. and Hassouna, A.F.A. (1996) Distribution of Some Heavy Metals in Qarounlake and River Nile, Egypt, Menofiya. Journal of Agricultural Research, 21, 733-746.

10. Dixon, W. and Chiswell, B. (1996) Review of Aquatic Monitoring Program Design. Water Research, 30, 1935-1948.

11. Reghunath, R., Murthy, T.R.S. and Raghavan, B.R. (2002) The Utility of Multivariate Statistical Techniques in Hy- drogeochemical Studies: An Example from Karnataka, India. Water Research, 36, 2437-2442.

12. McNeil, V.H., Coxa, M.E. and Preda, M. (2005) Assessment of Chemical Water Types and Their Spatial Variation Using Multi-Stage Cluster Analysis, Queensland, Australia. Journal of Hydrology, 310, 181-200.

13. Hansen, A.M., Leon-Zavala, A. and Bravo-Inclan, L. (1995) Sources of Pollution and Metal Enrichment in Sediments of the Lerma-Chapala Basin. Hydraulic Engineering in Mexico (in Spanish), 10, 55-69.

14. Trujillo-Cárdenas, J.L., Saucedo-Torres, N.P., Zárate del Valle, P.F., Ríos-Donato, N., Mendizábal, E. and Gómez- Salazar, S. (2010) Speciation and Sources of Toxic Metals in Sediments of Lake Chapala, Mexico. Journal of the Mexican Chemical Society, 54, 79-87.

15. Scott, C.A., Silva-Ochoa, P., Florencio-Cruz, V. and Wester, P. (2001) Competition for Water in the Lerma-Chapala Basin. Economic and Policy Implications of Water Transfers from Agricultural to Urban Uses. In: Hansen, A.M. and van Afferden, M., Eds., The Lerma-Chapala

Watershed: Evaluation and Management, Chapter 13, Kluwer Academic/Plenum Publishers, New York, 291-323.

16. Limon, J.G., Lind, O.T., Vodopich, D.S., Doyle, R. and Trotter, B.G. (1989) Long- and Short Term Variation in the Physical and Chemical Limnology of a Large, Shallow, Turbid Tropical Lake (Lake Chapala, Mexico). Archiv für Hydrobiologie, 83, 57-81.

17. De Anda, J. and Shear, H. (2001) Nutrients and Eutrophication in Lake Chapala. In: Hansen, A.M. and van Afferden, M., Eds., The Lerma-Chapala Watershed: Evaluation and Management, Kluwer Academic/Plenum Publishers, New York, 183-198.

18. Hansen, A.M. and van Afferden, M. (2001) Toxic Substances. Sources, Accumulation and Dynamics. In: Hansen, A.M. and van Afferden, M., Eds., The Lerma-Chapala Watershed: Evaluation and Management, Incluir no. de paginas, KluwerAcademic/PlenumPublishers, New York, 95-121.

19. NOM 001-SEMARNAT-1996. Diario Oficial de la Federacion. Norma Oficial Mexicana. Sets the Maximum Contaminant Limits in Wastewater Discharges to National Water Bodies. Mexico City. (In Spanish)

20. Rice, E.W., Baird, R.B., Eaton, A.D. and Clesceri, L.S. (2012) Standard Methods for the Examination of Water and Wastewater. 22nd Edition, APHA, AWWA, WEF, Baltimore.

21. De Bartolomeo, A., Poletti, L., Sanchini, G., Sebastiani, B. and Morozzi, G. (2004) Relationship among Parameters of Lake Polluted Sediments Evaluated by Multivariate Statistical Analysis. Chemosphere, 55, 1323-1329.

22. Altun, O., Türker Sacan, M. and Erdem, A.K. (2008) Water Quality and Heavy Metal Monitoring in Water and Sediment Samples of the Küçükçekmece Lagoon, Turkey. Environmental Monitoring and Assessment, 151, 345-362.

23. İşçen, C.F., Emiroglu, Ö., Ilhan, S., Arslan, N., Yilmaz, V. and Ahiska, S. (2008) Application of Multivariate Statistical Techniques in the Assessment of Surface Water Quality in Uluabat Lake, Turkey. Environmental Monitoring and Assessment, 144, 269-276.

24. Kotti, M.E., Vlessidis, A.G., Thanasoulias, N.C. and Evmiridis, N.P. (2005) Assessment of River Water Quality in Northwestern Greece. Water Resources Management, 19, 77-94.

25. Otto, M. (1998) Multivariate Methods. In: Kellner, R., Mermet, J.M., Otto, M. and Widmer, H.M., Eds., Analytical Chemistry, Wiley/VCH, Weinheim, 916.

26. McKenna Jr., J.E. (2003) An Enhanced Cluster Analysis Program with Bootstrap Significance Testing for Ecological Community Analysis. Environmental Modelling & Software, 18, 205-220.

27. WHO, World Health Organization (1997) Guidelines for Drinking Water Quality, Vol. 1, Recommendations. 2nd Edition, Geneva.

28. Abdul-Razak, A., Asiedu, A.B., Entsua-Mensah, R.E.M. and deGraft-Johnson, K.A.A. (2009) Assessment of the Water Quality of the Oti River in Ghana. West African Journal of Applied Ecology, 15, 1-12.

29. USEPA. National Recommended Water Quality Criteria. Aquatic Life Criteria Table and Human Health Criteria Table.http://water.epa.gov/scitech/swguidance/standards/current/index.cfm

Chapter 9

Moisture Control Methods in Silk Reeling Process of Tobacco Based on the Random Forest Regression

Bo Ma

School of Statistics and Mathematics, Yunnan University of Finance and Economics, Kunming, China

ABSTRACT

The moisture control of materials in silk reeling technology of tobacco is regarded as the key factor influencing the inner quality of cigarette. In this paper, according to the statistical data of the silk reeling production line of Yunyan (Ruanzhen brand) of Qujing cigarette factory from June 2013 to May 2014, it is feasible to apply the random forest regression model to study the problem of moisture control theoretically. In the perfuming stage of silk reeling, a random forest regression model is established to describe the change of moisture content of finished cut tobacco in the export link of perfuming stage, aroused by several factors including incoming water content and different environment. According to the model, good moisture control in the export link of perfuming stage (accordance with the technological standards) can be realized by adjusting the regulating reference value of incoming moisture under specific workshop environments. In the drying stage of silk reeling, the most effective method of moisture control is to adjust the cylinder wall temperature by means of analyzing the correlation coefficients among variables which influence the moisture content of cut tobacco in the export link of drying stage and then establishing another random forest regression model. And this method is consistent with the traditional production experiences. In conclusion, these methods referred above provide strong theoretical basis for stable moisture control in the export link of perfuming stage.

Keywords: Random Forests Regression, Moisture Control, Moisture Content in the Export Link of Drying Stage, Moisture Content in the Export Link of Perfuming Stage

1. INTRODUCTION

Silk reeling technology is an important section of cigarette production and the quality of it directly affects the working efficiency of the subsequent wrapping process. With the entire silk reeling process, from raw material preparing to tobacco cabinet-type storage, being companied by water, the material moisture control in the process is proved to be a key factor affecting the intrinsic quality of cigarettes. In more detail, good control of the material moisture will have many benefits in promoting the flexibility and process resistance of tobacco leaves, improving the structure of tobacco manufactures, increasing the output rate of silk and reducing the consumption of tobacco production.

Silk reeling, the longest section of cigarette production, is proved to be complicated. The production line is made up of five processes: loosing and moisture regaining, conditioning and casing, cutting, drying and perfuming. During the processes, tobacco leaves are changed into cut tobacco in leaf blade processing stage (the first three processes) and fed back into finished cut tobacco in the following drying and perfuming stage (the last two processes). Then, with finished cut tobacco being transferred into silos—storage cabinets, they are placed in different exit cabinets according to their brands and delivered to equipment in the cigarette packaging workshop, powered by wind through the specialized pipes. Furthermore, in order to realize the optimization management of the production process, the new silk reeling production line, whose layout is shown in Figure 1, is highlighted by their full use of automated control and monitoring and docking with enterprise management information system (MES).[1]

In silk reeling process, the existing moisture control targets focus on the moisture content of the finished cut tobacco in the export link of the perfuming stage which is supposed to accord with the technological standards. Due to the absence of moisture control function of perfuming machines, the goal of moisture control can be realized by means of adjustment of the moisture content of the cut tobacco in the export link of the drying stage, which has decisive influence on the control objective conversely. In actual production, workshop environment does affect the water desorption of tobacco during the transport, leading to the fluctuated gap between the moisture content in the export link of the drying stage and the perfuming stage. Therefore, it is so feasible to hunt the effective change rules of the gap referred above that we can adjust the moisture content in the export link of the drying stage in advance to guarantee the stable moisture content in the export link of the perfuming stage.

In silk reeling process, the changes of seasons and weather will cause changes in temperature and humidity of the workshop environment which will influence the moisture content of cut tobacco further. Consequently, in order to avoid the impacts brought by the factors above, it is feasible to timely adjust equipment parameters to control the moisture content of cut tobacco in the export link of the drying stage. Unfortunately, the frequent adjustments of equipment parameters completely depend on the production experience now, leading to the unstable gap of moisture content of cut tobacco in drying and perfuming stage. Aiming at making up for the deficiencies, with the appliance of the method of random forest regression (RFR) according to the data in one year, the model for hunting the variation law of water dissipation from the drying stage to the perfuming stage has been established. Considering the specific production environment when the moisture content of cut tobacco in the export link of the perfuming stage accords with the technological standards, we firstly use the model to predict the regulating reference value of moisture content in the export link of the dying stage. Then, we should find the varying adjustment ways of drying stage to keep the moisture content consistent with the regulating value. In the final step, based on the regulating reference value and adjustment ways, we can adjust equipment parameters of drying stage to the level where the gap of moisture content of cut tobacco in drying and perfuming stage is proved to be stable.

2. THE RANDOM FOREST ALGORITHM

Random forest algorithm is one kind of statistical theories, first used by Breiman in 2001 [1] . Based on the bootstrap re-sampling technology, this model extracts multiple samples from the original samples first, builds the decision tree models for every sample collection and combines the forecast of these decision tree models then, and votes for the final forecasting results.

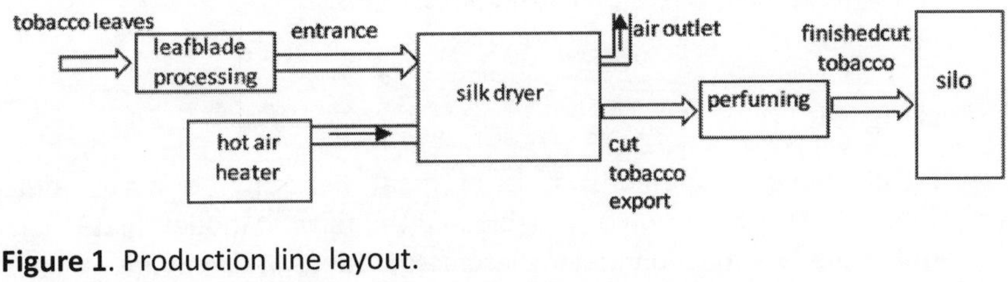

Figure 1. Production line layout.

2.1. Principle

Random forest regression, an important application of the random forest algorithm, is combined models consist of many decision tree models $\{h(X,\Theta_K),k=1,\cdots\}$, whose parameters set $\{\Theta_k\}$ is proved to be an independent and identically distributed random vector. Given an independent variable x, average prediction value $h(X,\Theta_k)$ under the decision tree models determines the final prediction value for each k. And the principle of random forest regression is shown in Figure 2[2] .

2.2. Features

In recent years, the random forest algorithm is more and more mature in both theories and methods. It has been widely applied in medicine, management, economics, and many other fields. And the superiority of the random forest algorithm embodied in many research results.

In the paper, Fang Kuangnan points out the following characteristics of random forest algorithm.

1) Compared with the AdaBoost algorithm, the accuracy of the random forest algorithm is comparable or even better. But, when referred to the computing speed, random forests algorithm has more significant advantages than AdaBoost algorithm [3] . And another advantage of the random forest algorithm is that the over-fitting phenomenon is not easy to produce.

2) The random forest algorithm can deal with continuous and categorical variables at the same time.

3) The random forest algorithm can also provide the useful information, such as internal error estimation, intensity, correlation coefficients, variable importance and others.

Considering the above advantages and the contrast of many machine learning methods (decision tree regression, Boosting regression, Bagging regression, artificial neural network regression etc.), it is obvious that the random forest algorithm has many advantages than other algorithms.

2.3. Testing

In traditional statistical methods, it is of great necessity to make some assumptions for the distribution of variable and the form of models first. Then, we determine the loss function under these assumptions and get various kinds of discrimination guidelines, including varying kinds of test standards and the critical value of statistics. However, it cannot be ignored that if the assumptions fail, these discrimination guidelines will not make any sense. Without the assumptions like traditional statistical methods, the quality of algorithm models is easy to be judged by cross-validation method—taking a part of the data as the training set to get the model, and then taking another part of the data as a test set to calculate the error. Therefore, considering the view of model assumptions, the algorithm models have a huge advantage and convenience compared with traditional statistical models.

In this paper, the model error can be calculated by the normalized mean square error (NMSE). And the normalized mean square error (NMSE) is defined as [4] :

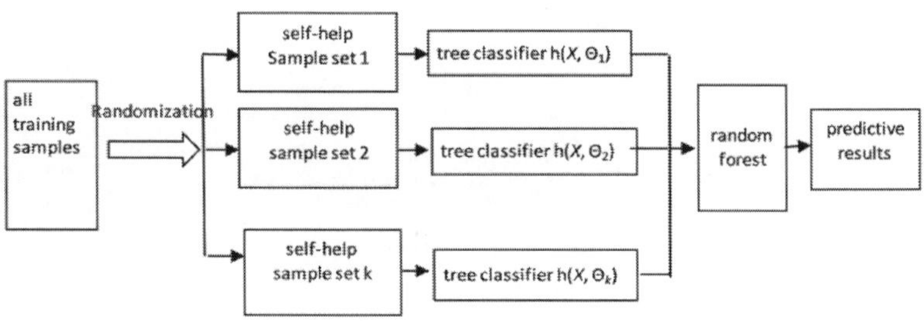

Figure 2. Random forest regression Sketch Map.

$$\text{NMSE} = \overline{(y-\hat{y})^2} / \overline{(y-\bar{y})^2} = \frac{\sum (y-\hat{y})^2}{\sum (y-\bar{y})^2}$$

In this formula, y represents the actual value of dependent variable of the test set and $\hat{y}$ represents its predictive value. It is obvious that if the model uses only mean value to predict, the normalized mean square error (NMSE) should be equal to 1. Consequently, if the NMSE of algorithm regression models is greater than 1, these models are proved to be very bad. And for the algorithm models, the smaller NMSE is the better (NMSE ≥ 0).

3. APPLICATION EXAMPLES

3.1. Data Sources and Analysis

In this paper, the statistical data of the silk reeling production line of Yunyan (Ruanzhen brand) of Qujing cigarette factory (from June 2013 to May 2014) are studied. And the statistical data come from enterprise management information system (MES) of Qujing cigarette factory.

We first observe and collect the data as Table 1 shown. In this table, every stage of silk reeling is divided into different processes which are defined and named. Then, after preliminary sorting, screening and matching with their codes, 1108 data samples of each processes of silk reeling of Yunyan (Ruanzhen brand) are obtained.

Shapiro-Wilk method is used to normality test of the original data, the results about the data are shown in Table 2.

Table 1. Definition of variables.

	Code	Batch Number
Tobacco Cutting	MCQS	The moisture content of tobacco cutting
Hot air Heater Parameters	THSRFF THSTB HSRFFL	Hot air temperature of drying stage Cylinder wall temperature of drying stage Hot air volume of drying stage
Air Outlet	MCHSCK THSCK THS HHS	Moisture content in the export link of the drying stage Temperature in the export link of the drying stage Temperature in drying stage Humidity in drying stage
Winnowing-Cooling	MCFXLQ	Moisture content by winnowing-cooling
Perfuming Export	MCJX, SMCJX	Moisture content in the export of the perfuming stage

Table 2. Results of Shapiro-Wilk normality test of all mean figures.

NO	Variable	W value	p-value	Variable	W value	p-value
1	MCQS	0.9971	0.03647	Ln(MCQS)	0.9967	0.01657
2	MCHSCK	0.979	7.84E−12	Ln(MCHSCK)	0.9804	2.56E−11
3	MCFXLQ	0.9695	8.58E−15	Ln(MCFXLQ)	0.9657	8.83E−16
4	MCJX	0.9815	6.56E−11	Ln(MCJX)	0.9824	1.54E−10
5	THC	0.976	7.91E−13	Ln(THC)	0.9572	<2.20E−16
6	HHC	0.9758	6.86E−13	Ln(HHC)	0.9247	<2.20E−16
7	TJL2	0.9698	1.03E−14	Ln(TJL2)	0.9523	<2.20E−16
8	HJL2	0.9501	<2.2E−16	Ln(HJL2)	0.9011	<2.20E−16
9	THS	0.965	5.58E−16	Ln(THS)	0.9509	<2.20E−16
10	HHS	0.9863	7.11E−09	Ln(HHS)	0.9479	<2.20E−16

The results show that under 5% of significance level, both the moisture data and the temperature-humidity data of each process refuse to obey the normal distribution assumption, even if the data was treated logarithmically. Therefore, the data are termed as non-normal distribution statistical data.

Owing to the obvious truth referred above, we should avoid the use of traditional statistical analysis methods based on normal distribution assumption and choose non-parameter statistics methods which do not depend on the data distribution. Accordingly, in this study, random forest regression analysis is feasible.

3.2. Application

In silk reeling process, the existing moisture control targets focus on the moisture content of the finished cut tobacco in the export link of the perfuming stage which is supposed to accord with the technological standards. Therefore, we should first establish the model for hunting the variation law of water dissipation from the drying stage to the perfuming stage. Then, we should find the varying adjustment ways of drying stage to control the moisture content in actual production.

3.2.1. Perfuming Stage

In perfuming stage, under the specific environment, the moisture content of cut tobacco in the export link of the drying stage proves to be the focus of the study, for the reason that we can use the data to calculate the moisture content in the export link of the perfuming stage and make a judgment whether the indicator accords with the technological standards. Research clues are as follows: first, we establish a model from the data samples, which are able to predict the corresponding moisture content in the export link of the perfuming stage from the drying stage; second, we can screen the suitable moisture content in the export link of the drying stage under the specific environment to satisfy the technological standards.

Theoretically, important variables which influence the moisture content in the export link of the perfuming stage include the moisture content and temperature in the export link of the drying stage and the temperature and humidity of the workshop environment. Hereby, we consider using one-year data to establish random forest regression model. In the model, the moisture content in the export link of the perfuming stage (MCJX) is to be as a dependent variable; the moisture content (MCHSCK) and the temperature (THSCK) in the export link of the drying stage and the temperature (THS) and humidity (HHS) in the drying stage are to be as independent variables.

Owing to the large amount of calculation of the prediction made by random forest regression model, we apply the packages Random Forest of R software to be as an critical implement to fulfill the study on the moisture content of silk reeling processes in this paper.

We use the standardized mean square error (NMSE) to evaluate the random forest regression model, with randomly selected 50 samples from the finished samples as test set and the rest of the samples as the training set. By means of training and testing the corresponding set of samples of the random forest regression model, the result shows the NMSE = 0.1807012 and the variable importance diagram [5] is shown in Figure 3.

Furthermore, we can also consider the proportion of the temperature of cut tobacco in the export link of drying

stage to the temperature during drying stage as a variable, which proves to be more reasonable. $T = \frac{THSCK}{HHS}$,

an added variable to the model, is termed as the relative temperature in the export link of drying stage.

Then, we also consider establishing a new random forest regression model. In this new model, the moisture content in the export link of the perfuming stage (MCJX) is to be as a dependent variable; the moisture content (MCHSCK) in the export link of the drying stage, the humidity (HHS) in the drying stage and the relative temperature during the drying stage are to be as independent variables. The result shows the NMSE = 0.1828422 and the variable importance diagram is shown in Figure 4.

From the results of the analysis above, using the relative variable T, instead of THSCK and HHS, as an important independent variable which influence the moisture content in the export link of the perfuming stage, not only reduces the number of variables of the model, but enhances the accuracy of prediction.

According to the technological standards of the silk processes and the new random forest regression model, using the reconstructed data of all possible environmental temperature and humidity combinations and moisture content of cut tobacco in the export link of the drying stage under different circumstances, we can predict the corresponding moisture content in the export link of the perfuming stage.

Figure 5 shows the distribution of the moisture content of finished cut tobacco in the export link of perfuming stage.

It can be seen from the chart that the annual moisture content of finished cut tobacco in the export link of perfuming stage concentrates on the interval [12.15, 12.65]. Based on the interval, the moisture content in the export link of drying stage can be calculated.

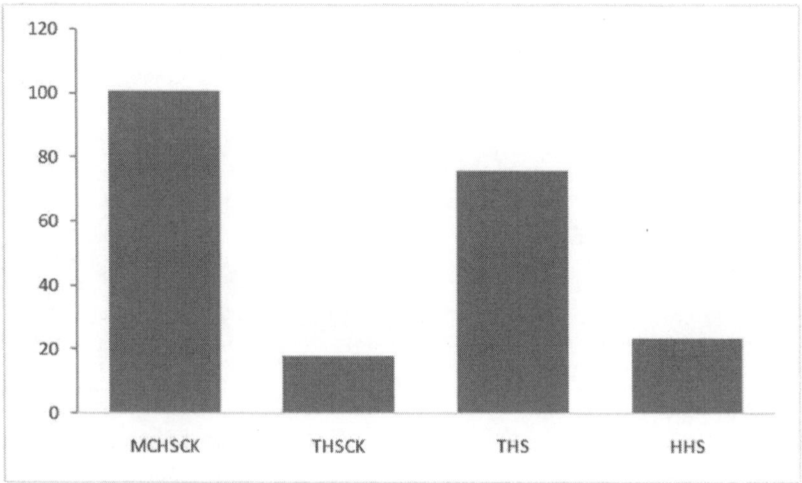

Figure 3. Variable importance diagram 1.

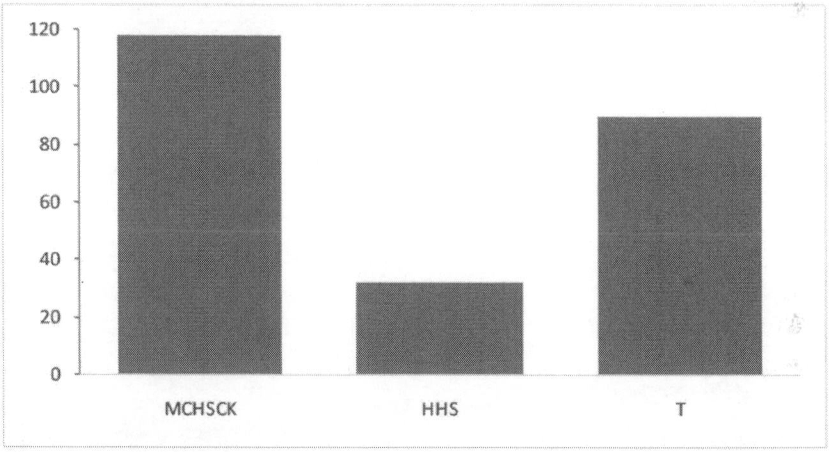

Figure 4. Variable importance diagram 2.

Then, we establish the prediction set to forecast the moisture content of finished cut tobacco in the export link of perfuming stage. And the distribution situation of all variables (MCHSCK, T, HHS) is shown in Figure 6 for intuitive understanding about the distribution range.

From the chart above, the moisture content of cut tobacco in the export link of drying stage (MCHSCK) concentrates on the interval [12.50, 13.50]; the distribution range of humidity in the drying stage (HHS) is [30, 55]; the distribution range of relative temperature in drying stage (T) is [1.65, 2.50]. According to the distribution situation of all variables, we choose MCHSCK 101 samples, HHS 26 samples and T 81 samples to establish the testing set, which

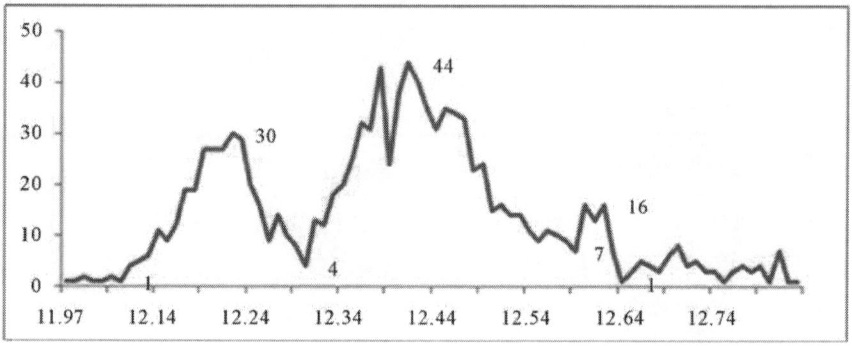

Figure 5. Distribution of the Moisture Content of Finished Cut Tobacco in the Export Link of Perfuming Stage.

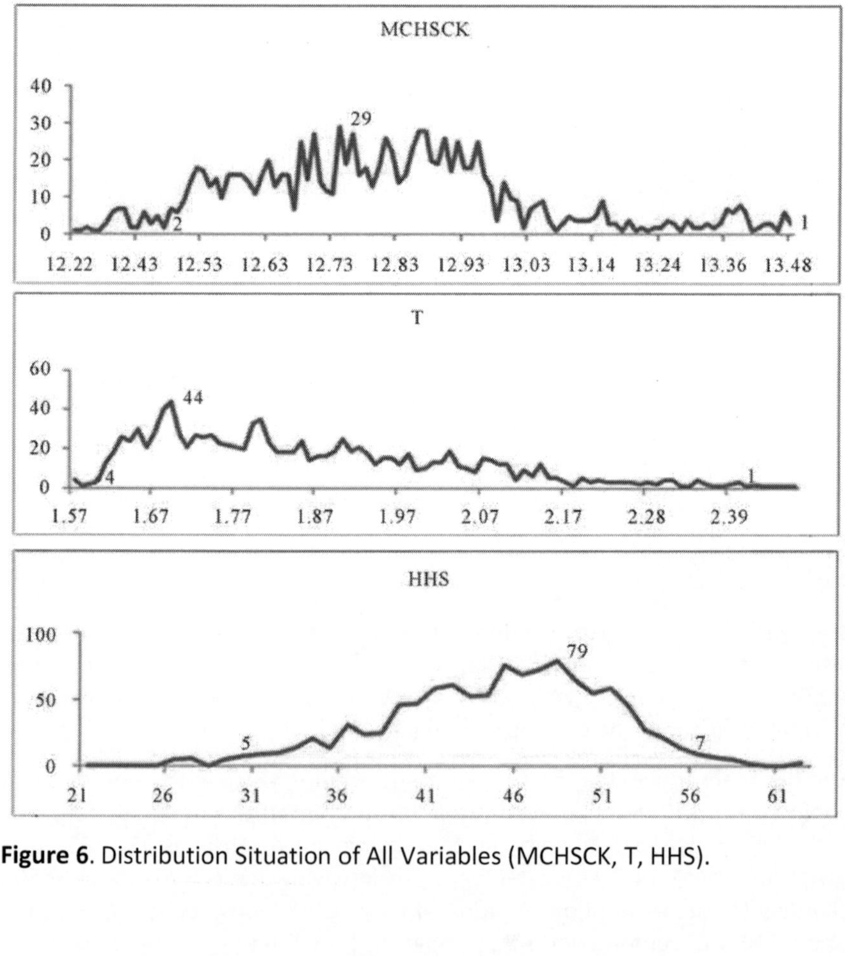

Figure 6. Distribution Situation of All Variables (MCHSCK, T, HHS).

has total 212706 samples. At the same time, we also establish the training set, including all samples of variables (MCHSCK, HHS, T and MCJX) from the original data, to do random forest regression, where MCJX is to be as the dependent variable, and do prediction with the constructed prediction set. The predictive values of moisture content in the export link of perfuming stage are listed in Table 3, when T value equals to 1.6; HHSJ value equals to 30; the moisture content in the export link of drying stage varies during the interval [12.5, 13.5].

Table 3. Parts of predictive results.

MCHSCK	Predictive Value	MCHSCK	Predictive Value	MCHSCK	Predictive Value	MCHSCK	Predictive Value
12.5	12.34617083	12.76	12.42491706	13.02	12.56127533	13.28	12.6565013
12.51	12.34942478	12.77	12.42864518	13.03	12.56455	13.29	12.6534287
12.52	12.35046444	12.78	12.42633967	13.04	12.57613567	13.3	12.6537
12.53	12.34828032	12.79	12.42708433	13.05	12.577577	13.31	12.648417
12.54	12.35662644	12.8	12.42699182	13.06	12.59246381	13.32	12.651027
12.55	12.35730533	12.81	12.44784695	13.07	12.59901148	13.33	12.64707933
12.56	12.35996767	12.82	12.44853633	13.08	12.65028467	13.34	12.64392367
12.57	12.36128724	12.83	12.45159261	13.09	12.655787	13.35	12.64354867
12.58	12.363723	12.84	12.45248242	13.1	12.66488433	13.36	12.627273
12.59	12.364025	12.85	12.45121233	13.11	12.66658267	13.37	12.62651562
12.6	12.3655927	12.86	12.45623567	13.12	12.66643467	13.38	12.62746317
12.61	12.36601467	12.87	12.45501567	13.13	12.66656333	13.39	12.6276655
12.62	12.36862217	12.88	12.455084	13.14	12.66678733	13.4	12.6221605
12.63	12.36792083	12.89	12.455004	13.15	12.66810293	13.41	12.62293
12.64	12.367584	12.9	12.47731656	13.16	12.66842505	13.42	12.63119167
12.65	12.36934726	12.91	12.47869833	13.17	12.66883933	13.43	12.63215967
12.66	12.37752714	12.92	12.48422067	13.18	12.66791633	13.44	12.632419
12.67	12.38122167	12.93	12.48870857	13.19	12.66755433	13.45	12.63278
12.68	12.39118277	12.94	12.50823	13.2	12.66775467	13.46	12.63278
12.69	12.41271179	12.95	12.51183878	13.21	12.66351033	13.47	12.63234833
12.7	12.40669296	12.96	12.51379767	13.22	12.66706133	13.48	12.63234833
12.71	12.40819267	12.97	12.51557629	13.23	12.66846167	13.49	12.63234833
12.72	12.40897467	12.98	12.52452267	13.24	12.67301833	13.5	12.63234833
12.73	12.42308314	12.99	12.53225733	13.25	12.67317967		
12.74	12.42327892	13	12.539555	13.26	12.672694		
12.75	12.42329995	13.01	12.55258333	13.27	12.66729333		

In Table 3, when the moisture content in the export link of drying stage equals to 12.94; T value equals to 1.6; HHSJ value equals to 30, the predictive value of moisture content in the export link of perfuming stage is 12.50823 which is the best figure according to the technical standards 12.5. Therefore, the regulating reference value of moisture content in the export link of drying stage is 12.94.

By means of the same method, the regulating reference of moisture content in the export link of drying stage can be calculated under different environments.

3.2.2. Drying Stage

The moisture content of cut tobacco in the export link of drying stage is a controllable variable, affected by setting parameters of drying machines. In this stage, moisture control is the core problem studied, namely how to adjust the setting parameters of drying machines to control the moisture content. Research clues are as follows: first research focus on the correlation between all parameters of drying machines and the moisture content in the export link of drying stage; then, analyzing the importance of all variables which influence the moisture content. According to this analysis, we can adjust the important variables of drying stage in this stage.

Production experiences show that the factors which influence the moisture content in the export link of drying stage include the incoming water content and setting parameters of drying machines. The incoming water refers to the moisture content of tobacco cutting (MCQS). Setting parameters of drying machines include the hot air temperature (THSRF), cylinder wall temperature (THSTB) and hot air volume (HSREFL) of drying stage. Considering these variables above, when the data of them do not obey the normal distribution assumption, it is feasible to use Spearman correlation coefficient analysis method to describe the correlation among variables.

By means of software R, the Spearman correlation coefficient between the moisture content in the export link of drying stage and other four variables in Table 4.

Table 4 shows a positive correlation between the moisture content in the export link of drying stage (MCHSCK) and the moisture content of tobacco cutting (MCQS); inverse correlations between the moisture content in the export link of drying stage (MCHSCK) and the hot air temperature (THSRF), cylinder wall temperature (THSTB) and hot air volume (HSREFL) of drying stage respectively. According to the above correlations between MCHSCK and MCQS, THSRF, THSTB and HSREFL, the moisture content in the export link of drying stage can be controlled by means of adjusting the value of these four variables. A random forest regression model is supposed to be established to study the moisture content in the export link of drying stage and the adjustment can be carried out according to the importance of four variables referred above.

Table 4. Spearman correlation coefficient.

	MCQS	THSRF	THSTB	HSRFFL
MCHSCK	0.353325	−0.29472	−0.29969	−0.31212

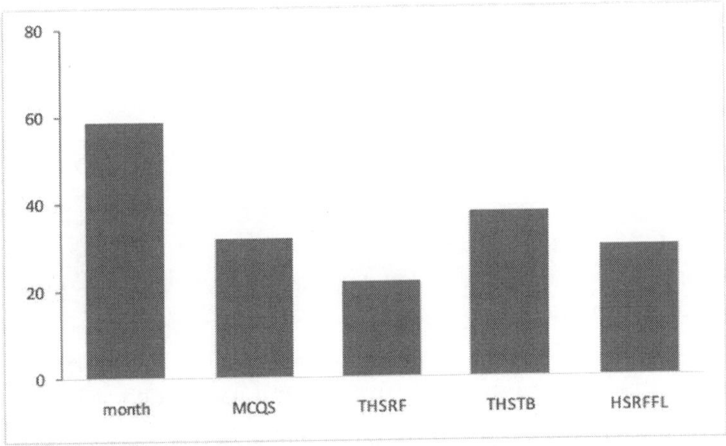

Figure 7. Variable importance diagram 3.

Hereby, a random forest model has been established with the moisture content in the export link of drying stage (MCHSCK) as the dependent variable; months, the hot air temperature (THSRF), cylinder wall temperature (THSTB) and hot air volume (HSREFL) of drying stage as the independent variables. And the model results show that the standardized mean square error (NMSE) equals to 0.271858; the variable THSTB is proved to be the most important variable fromFigure 7. Therefore, the most effective measure to control the moisture content in the export link of drying stage is to adjust the cylinder wall temperature. It is understood that the actual production emphasizes the adjustment of the cylinder wall temperature along with the adjustment of the hot air volume, which is consistent with the theoretical analysis results above.

4. CONCLUSIONS

In silk reeling process of tobacco, moisture control is of great significance to improving the quality of cigarettes. Due to the characteristics of nonlinearity and uncertainty of the data in the drying stage of silk reeling, in this paper, the random forest algorithm has been used to simulate the reconstructed prediction set in order to find the reference standards of the moisture content in the export of drying stage. Therefore, the moisture content in the export of perfuming stage can be controlled according the reference standards referred above.

It is well-known that traditional multiple linear regression cannot deal with non-normal distribution data, but the random forest regression model effectively compensates for this defect [6] . Furthermore, this model has the advantages of high accuracy, high extrapolation ability and less adjustment parameters.

REFERENCES

1. Breiman, L. (2001) Random Forests. Machine Learning, 45, 5-32.
2. Svetnik, V., Liaw, A., Tong, C., Culberson, J.C., Sheridan, R.P. and Feuston, B.P. (2003) Random Forest: A Classification and Regression Tool for Compound Classification and QSAR Modeling. Journal of Chemical Information and Computer Sciences, 43, 1947-1958.
3. Lin, Y. and Jeon, Y. (2006) Random Forests and Adaptive Nearest Neighbors. Journal of the American Statistical Association, 101, 578-590.
4. Wu, X.Z. (2013) Statistical Methods for Complex Data—Based on R. China Renmin University Press Co., Ltd, Beijing, 33.
5. van der Laan, M.J. (2006) Statistical Inference for Variable Importance. International Journal of Biostatistics, 2, Article 2.
6. Peters, J., De Baets, B., Verhoest, N.E.C., Samson, R., Degroeve, S., De Becker, P. and Huybrechts, W. (2007) Random Forests as a Tool for Ecohydrological Distribution Modelling. Ecological Modelling, 207, 304-318.

NOTES

[1]MES stems from the silk reeling production line of Yunyan (Ruanzhen brand) of Qujing cigarette factory.

Chapter 10

One-Sample Bayesian Predictive Analyses for an Exponential Non-Homogeneous Poisson Process in Software Reliability

Albert Orwa Akuno, Luke Akong'o Orawo, Ali Salim Islam

Department of Mathematics, Egerton University, Egerton, Kenya

ABSTRACT

The Goel-Okumoto software reliability model, also known as the Exponential Nonhomogeneous Poisson Process, is one of the earliest software reliability models to be proposed. From literature, it is evident that most of the study that has been done on the Goel-Okumoto software reliability model is parameter estimation using the MLE method and model fit. It is widely known that predictive analysis is very useful for modifying, debugging and determining when to terminate software development testing process. However, there is a conspicuous absence of literature on both the classical and Bayesian predictive analyses on the model. This paper presents some results about predictive analyses for the Goel-Okumoto software reliability model. Driven by the requirement of highly reliable software used in computers embedded in automotive, mechanical and safety control systems, industrial and quality process control, real-time sensor networks, aircrafts, nuclear reactors among others, we address four issues in single-sample prediction associated closely with software development process. We have adopted Bayesian methods based on non-informative priors to develop explicit solutions to these problems. An example with real data in the form of time between software failures will be used to illustrate the developed methodologies.

Keywords: Nonhomogeneous Poisson Process, Non-Informative Priors, Software Reliability Models, Bayesian Approach

1. INTRODUCTION

Over the last decade of the 20th century and the first few years of the 21st century, the demand for complex software systems has gone high as it is seen that today, computers are embedded in automotive mechanical and safety control systems, industrial and quality process control, real-time sensor networks, aircrafts, nuclear reactors, hospital healthcare and air traffic control systems among others; computer systems have become an indispensable component of our modern society today. Consequently, the reliability of software used in these systems has been a major concern and a requirement in the modern generation. Software reliability is defined as the probability of failure-free software operations for a specified period of time in a specified environment [1] . A single software defect can cause system failure and to avoid these failures, reliable software is required. Software reliability is achieved through testing during the software development testing stage [2] . The usual criteria of removing bugs in software are by running test cases in a manner that exercises the software similar to the way that users will operate in their particular environment. However, emulating end-user environment during the test interval is difficult and time-consuming especially when there are multiple types of end-users and also, business pressure to release a software system within a tight market window puts a constraint on the amount of time that can be spent testing the software. Software reliability modeling comes in handy to address this dilemma. As indicated by [3] , software reliability modeling can provide the basis for planning reliability growth tests, monitoring progress and estimating current reliability and forecasting and predicting future reliability improvements. Forecasting and prediction are achieved through predictive analyses. In particular, predictive analyses are useful in determining when to terminate the development process of software or hardware. Often, a prediction interval is constructed to provide the time frame when the $k\text{th}\,(k>0)$ future failure observation will occur with a pre-determined confidence level [4] .

Many software reliability models have been developed by various authors and researchers in the past three decades. Amongst, an Exponential Nonhomogeneous Poisson Process with intensity function

$$\lambda(t)=\alpha\beta e^{-\beta t} \tag{1}$$

is the earliest software reliability model to be developed by Goel and Okumoto in 1979. In various literatures, this NHPP is called the Goel-Okumoto (1979) model.

As noted by [5] , the Goel-Okumoto (1979) model has been applied to a number of software testing environments and its application on assessing and detecting software failures has been investigated by various authors. For instance, the Goel-Okumoto model has been used to develop a statistical control mechanism that could be used to detect whether a software process is statistically under control or not. ML estimation of the parameters of the Goel-Okumoto (1979) model has been conducted and in particular, it has been shown that the ML estimates of the parameters of the model are not consistent as the testing period extends to infinity. [6] presented an empirical method for selecting software reliability growth models for release decision-making where they applied iteratively various software reliability models namely Goel-Okumoto (1979), Delayed S-shaped, Gompertz and Yamada exponential software reliability growth models to weekly cumulative software failure data during system test to determine the number of remaining failures expected in software after release. [7] also performed parameter estimation of the Goel-Okumoto, Yamada S-shaped and Inflection S-shaped software reliability growth models where they also established a necessary and sufficient condition with respect to the software failure data, of which, if satisfied, will ensure that the MLE method returns a unique positive and finite estimation of the unknown parameters of the Goel-Okumoto and the Yamada S-shaped models. [8] presented software failure data which, after study, depicted that the failure rate, i.e. the number of failures per hour, seemed to be decreasing with time, an indication that a Nonhomogeneous Poisson Process with mean value function $m(t)= \alpha\left(1-e^{-\beta t}\right)$, a mean value function corresponding to that of the Goel-Okumoto software reliability model, was a reasonable model to describe the failure process. From the literature, it is evident that most of the study that has been done on the Goel-Okumoto software reliability model is parameter estimation using, especially, the MLE method and model fit. There is a conspicuous absence of literature on both the classical and Bayesian predictive analyses on the model.

This paper focuses on single-sample predictive inference for the Goel-Okumoto (1979) software reliability model using Bayesian approach. We first identify four issues in the single-sample prediction associated closely with the development testing process of software and proceed to develop and derive the corresponding predictive distributions in Section 2. The main results for single-sample prediction are presented in Section 3. A real example in the form of secondary software failure data in the form of execution times between successive software failures is used to illustrate the proposed and developed methodologies in Section 4. A discussion is given in Section 5 and thereafter, mathematical proofs are given in the Appendix.

2. PREDICTIVE ISSUES AND BAYESIAN METHOD

During the development testing stage of a software, statisticians and engineers are overly interested in various predictive problems whose solutions are believed to be very important in modifying, debugging and determining when to terminate software development testing process. In this section, we present four issues associated closely with software development testing process and derive the predictive distributions using Bayesian approach. For the purposes of the four predictive issues, we assume that a reliability growth testing is performed on a software and the cumulative number of failures of the software in the time interval $(0,t]$, denoted by $N(t)$ is observed. We further assume that $[N(t), t>0]$ follows the NHPP with intensity function given in Equation (1). Let $0<t_1<t_2<\cdots$ be the observed failure times. Failure data is said to be failure-truncated when testing stops after a predetermined n number of failures occur. We denote the n failure times by $Y_{obs}^{f}=[t_i]_{i=1}^{n}$ where $0<t_1<t_2<\cdots<t_n$. Failure data is said to be time truncated if testing stops at a predetermined time t. We denote the corresponding observed failure data by $Y_{obs}^{t}=[n,t_1,\cdots,t_n;t]$, where $0<t_1<t_2<\cdots<t_n\le t$.

Prediction interval is a confidence interval for a future observation or a function of some future observations. Specifically, a double-sided (bilateral) prediction interval for t_{n+k} with confidence level γ is defined by $\left[T_{n+k,l(\gamma)}, T_{n+k,u(\gamma)}\right]$ such that $\Pr\left[T_{n+k,l(\gamma)}\le t_{n+k}\le T_{n+k,u(\gamma)}\right]=\gamma$. Similarly, a single-sided (unilateral) lower or upper prediction limit for t_{n+k} with level γ is defined by $T_{n+k,L(\gamma)}$ (or $T_{n+k,U(\gamma)}$) which satisfies $\Pr\left[T_{n+k,L(\gamma)}\le t_{n+k}\right]=\gamma$ (or $\Pr\left[t_{n+k}\le T_{n+k,U(\gamma)}\right]=\gamma$).

Both $T_{n+k,L(\gamma)}$ and $T_{n+k,U(\gamma)}$ depends only on a single sample (or a single software) and are called single-sample prediction limits. Prediction limits involving two samples (or two softwares) can be defined similarly and are called two-sample prediction limits.

2.1. Issues in Single-Sample Software Reliability Prediction

Here, we consider one software and assume that its cumulative inter-failure times obey the Goel-Okumoto (1979) software reliability model with observed data being either Y_{obs}^{f} or Y_{obs}^{t}. Based on Y_{obs}^{f} or Y_{obs}^{t}, we are interested in the following problems:

Issue A: what is the probability that at most k software failures will occur in the future time period $(T,\tau]$ with $\tau > T$?

Issue B: suppose that the pre-determined target value λ_{tv} for the failure rate of the software undergoing development testing is not achieved at time T, what is the probability that the target value λ_{tv} will be achieved at time $\tau, \tau > T$?

Issue C: suppose that the target value λ_{tv} for the software failure rate is not achieved at time T, how long will it take so that the software failure rate will be attained at λ_{tv}?

Issue D: what is the upper prediction limit (UPL) of $\lambda_{\tau} = \alpha\beta e^{\beta\tau}$ with level γ, τ being a predetermined value greater than T?

2.2. Posterior and Predictive Distributions

Let Y_{obs} represent Y_{obs}^{f} or Y_{obs}^{t}. The joint density of Y_{obs} is therefore

$$f\left(Y_{obs}\middle|\alpha,\beta\right) = \alpha^{n}\beta^{n} e^{-\beta\sum_{i=1}^{n} t_i} e^{-\alpha\left(1-e^{-\beta T}\right)} \tag{2}$$

Case 1: When the shape parameter β is known, we adopt the following non-informative prior distribution of α:

$$\pi(\alpha) \propto \frac{1}{\alpha}, \ \alpha > 0$$.(3)

The posterior distribution of α is thus given by

$$h(\alpha|Y_{obs}) = [\Gamma(n)]^{-1} \alpha^{n-1} e^{-\alpha(1-e^{-\beta T})} (1-e^{-\beta T})^{n}$$ (4)

Let y^{+} be the random variable being predicted. Then the posterior predictive distribution of y^{+} is give as

$$f(y^{+}|Y_{obs}) = \int_{0}^{\infty} f(y^{+}|Y_{obs}, \alpha) h(\alpha|Y_{obs}) d\alpha$$ (5)

Hence the Bayesian UPL of y^{+} with level γ denoted as $y_{U}^{(\beta)}$ must satisfy

$$\gamma = \int_{-\infty}^{y_{U}^{(\beta)}} f(y^{+}|Y_{obs}) dy^{+}$$.(6)

Case 2: When the shape parameter β is unknown, we consider the following non-informative joint prior density for α and β (we assume that α and β are independent).

$$\pi(\alpha, \beta) \propto \frac{1}{\alpha\beta}, \quad \alpha, \beta > 0.$$ (7)

Hence the corresponding joint posterior density is given as

$$h\left(\alpha,\beta \middle| Y_{obs}\right)=\left[k\Gamma(n)\right]^{-1}\alpha^{n-1}\beta^{n-1}\mathrm{e}^{-\beta\sum_{i=1}^{n}t_i}\mathrm{e}^{-\alpha\left(1-\mathrm{e}^{-\beta T}\right)} \quad (8)$$

where

$$k=\int_0^\infty \frac{\beta^{n-1}\mathrm{e}^{-\beta\sum_{i=1}^{n}t_i}}{\left(1-\mathrm{e}^{-\beta T}\right)^n}\mathrm{d}\beta \quad .(9)$$

Similar to Equation (5) and Equation (6), let yU denote the Bayesian UPL of y with level γ, then

$$f\left(y^+ \middle| Y_{obs}\right)=\int_0^\infty\int_0^\infty f\left(y^+ \middle| Y_{obs},\alpha,\beta\right)h\left(\alpha,\beta \middle| Y_{obs}\right)\mathrm{d}\alpha\mathrm{d}\beta \quad (10)$$

And

$$\gamma=\int_{-\infty}^{yU} f\left(y^+ \middle| Y_{obs}\right)\mathrm{d}y^+ \quad (11)$$

3. MAIN RESULTS FOR THE PREDICTION PROBLEMS

In this section, we address the four single-sample prediction issues raised in Section 2.1 using Bayesian approach. The following propositions are considered as the main results with proofs being given in the Appendix. In the subsequent results, we use $\chi^2(n;\gamma)$ to represent the γ percentage point of the chi-square distribution with n degrees of freedom and we also assume the priors to be Equation (3) and Equation (7).

Proposition 1 (for issue A): The probability that at most k software failures will occur in the future time period $(T,\tau]$ with $\tau > T$ is

$$\gamma_k = \begin{cases} \left(\dfrac{1-e^{-\beta T}}{e^{-\beta T}-e^{-\beta\tau}}\right)^n \displaystyle\sum_{j=n}^{n+k}\binom{j-1}{n-1}\left(\dfrac{e^{-\beta T}-e^{-\beta\tau}}{1-e^{-\beta\tau}}\right)^j, & \text{when } \beta \text{ is known} \\ \displaystyle\sum_{j=n}^{n+k}\dfrac{\Gamma(n)}{d(j-n)!\Gamma(n)}\int_0^\infty \beta^{n-1} e^{-\beta\sum_{i=1}^n t_i} \dfrac{\left(e^{-\beta T}-e^{-\beta\tau}\right)^{j-n}}{\left(1-e^{-\beta\tau}\right)^j} d\beta, & \text{when } \beta \text{ is unknown} \end{cases} \quad (12)$$

Proposition 2 (for issue B): Suppose that the pre-determined target value λ_{tv} for the failure rate of the software undergoing development testing is not achieved at time T, the probability that the target value λ_{tv} will be achieved at time $\tau, \tau > T$ is

$$\gamma = \begin{cases} 1-\displaystyle\sum_{h=0}^{n-1}\dfrac{\left(\dfrac{1-e^{-\beta T}}{\beta e^{-\beta\tau}}\lambda_{tv}\right)^h}{h!} e^{-\frac{1-e^{-\beta T}}{\beta e^{-\beta\tau}}\lambda_{tv}} & \text{if } \beta \text{ is known} \\ 1-\dfrac{1}{k}\displaystyle\sum_{h=0}^{n-1}\int_0^\infty \dfrac{\left(\dfrac{1-e^{-\beta T}}{\beta e^{-\beta\tau}}\lambda_{tv}\right)^h}{h!} e^{-\frac{\left(1-e^{-\beta T}\right)}{\beta e^{-\beta\tau}}\lambda_{tv}} \dfrac{\beta^{n-1} e^{-\beta\sum_{i=1}^n t_i}}{\left(1-e^{-\beta T}\right)^n} d\beta & \text{if } \beta \text{ is unknown} \end{cases} \quad (13)$$

Remark 1: Let $\left\{\beta^{(\ell)}\right\}_{\ell=1}^{L}$ be i.i.d. sample from $\mathrm{Gamma}\left(n, \sum_{i=1}^n t_i\right)$, we can approximate the second part of (13) via MCMC method.

Proposition 3 (for issue C): For given level γ, the time τ^* required to attain λ_{tv} is

(i) $$\tau^* = \frac{1}{\beta}\ln\left[\frac{\beta\chi^2(2n;\gamma)}{2\lambda_{tv}\left(1-e^{-\beta T}\right)}\right] - T \quad \text{if } \beta \text{ is known} \quad (14)$$

(ii) $$\tau^* = \tau - T \quad \text{if } \beta \text{ is unknown} \quad (15)$$

where τ is the solution to the following equation:

$$\gamma = 1 - \frac{1}{k}\sum_{h=0}^{n-1}\int_0^{\infty} \frac{\left(\frac{1-\mathrm{e}^{-\beta T}}{\beta \mathrm{e}^{-\beta\tau}}\lambda_{tv}\right)^h}{h!} \mathrm{e}^{-\frac{\left(1-\mathrm{e}^{-\beta T}\right)}{\beta \mathrm{e}^{-\beta\tau}}\lambda_{tv}} \frac{\beta^{n-1}\mathrm{e}^{-\beta\sum_{i=1}^{n} t_i}}{\left(1-\mathrm{e}^{-\beta T}\right)^n} \mathrm{d}\beta \tag{16}$$

Proposition 4 (for issue D): The Bayesian UPL of $\lambda_\tau = \alpha\beta \mathrm{e}^{\beta\tau}$ with level γ is

(i)

$$\lambda_U^{(\beta)} = \frac{\chi^2(2n;\gamma)\beta \mathrm{e}^{-\beta\tau}}{2\left(1-\mathrm{e}^{-\beta T}\right)} \quad \text{if } \beta \text{ is known} \tag{17}$$

(ii) (β unknown) λ_{tv} such that (18)

$$\gamma = 1 - \frac{1}{k}\sum_{h=0}^{n-1}\int_0^{\infty} \frac{\left(\frac{1-\mathrm{e}^{-\beta T}}{\beta \mathrm{e}^{-\beta\tau}}\lambda_{tv}\right)^h}{h!} \mathrm{e}^{-\frac{\left(1-\mathrm{e}^{-\beta T}\right)}{\beta \mathrm{e}^{-\beta\tau}}\lambda_{tv}} \frac{\beta^{n-1}\mathrm{e}^{-\beta\sum_{i=1}^{n} t_i}}{\left(1-\mathrm{e}^{-\beta T}\right)^n} \mathrm{d}\beta \tag{19}$$

4. EXAMPLE

In this section, a real example from the time between failure data given by [9] is used to illustrate the developed methodologies for the single-sample Bayesian predictive analysis. The Table 1 gives the Time Between Failure.

The study has used the cumulative time between failures as failure times $0 < t_1 < t_2 < \cdots < t_{30}$ where $n = 30$. These data obey the Goel-Okumoto (1979) software reliability model [10] . The MLEs of the parameters of the software reliability model based on the data are $\hat{\alpha} = 31.698171$ and $\hat{\beta} = 0.003962$. In the illustration of the developed methodologies, the study has used these MLEs.

1) Suppose that we are interested in the probability γ_k that at most k failures will occur in the future time period $(\tau, T] = (180, 250]$. a) When β is known (say $\beta = 0.003969$), using the first formula in Equation (12), we have $\gamma_0 = 0.0039$, $\gamma_1 = 0.0235$, $\gamma_2 = 0.0750$, $\gamma_3 = 0.1677$, $\gamma_4 = 0.2970$, $\gamma_5 = 0.4456$, $\gamma_6 = 0.5920$, $\gamma_7 =$ 0.7193, $\gamma_8 = 0.8188$, $\gamma_9 = 0.8898$, $\gamma_{10} = 0.9366$, $\gamma_{11} = 0.9653$, $\gamma_{12} = 0.9819$, $\gamma_{13} = 0.9910$, $\gamma_{14} = 0.9957,$ and $\gamma_{15} = 0.9980.$ b) When β is unknown, from the second formula in Equation (12) we obtain, $\gamma_0 = 0.0010$, $\gamma_1 = 0.0087$, $\gamma_2 = 0.0292$, $\gamma_3 = 0.0721$, $\gamma_4 = 0.1575$, $\gamma_5 = 0.2566$, $\gamma_6 = 0.3920$, $\gamma_7 = 0.5099$, $\gamma_8 = 0.6344$, $\gamma_9 = 0.7534$, $\gamma_{10} = 0.8350$, $\gamma_{11} = 0.9062$, $\gamma_{12} = 0.9462$, $\gamma_{13} = 0.9621$, $\gamma_{14} = 0.9831$, $\gamma_{15} = 0.9944$.

Figure 1 shows the graph of the desired probabilities for the case when β is known and when β is unknown.

2) Suppose that the target value is given by $\lambda_{tv} = 0.03$. At time $T = 182.21$, the MLE of the achieved failure rate for this software is $\hat{\lambda}(182.21) = \hat{\alpha}\hat{\beta}e^{-182.21\hat{\beta}} = 0.061$ which is greater than λ_{tv} i.e. it cannot be achieved at time $T = 182.21$. Thus the development testing will continue. Suppose we want to predict the probability that the target value λ_{tv} will be achieved at time $\tau = 277.83$. a) When β is known, say $\beta = 0.003962$, from the first formula in Equation (13) we obtain $\gamma = 0$. Thus we can conclude that the target value (failure rate) will not be achieved. b) When β is unknown, from the second formula in Equation (13) and Remark 1, we obtain $\gamma = 0.0576$ where the Monte Carlo sample size is $L = 1000$.

Table 1. Time between failures data.

Failure No.	Time between failures	Cumulative time between failures	Failure No.	Time between failures	Cumulative time between failures
1	30.02	30.02	16	15.53	151.78
2	1.44	31.46	17	25.72	177.50
3	22.47	53.93	18	2.79	180.29
4	1.36	55.29	19	1.92	182.21
5	3.43	58.72	20	4.13	186.34
6	13.2	71.92	21	70.47	256.81
7	5.15	77.07	22	17.07	273.88
8	3.83	80.90	23	3.99	277.83
9	21	101.90	24	176.06	453.93
10	12.97	114.87	25	81.07	535.00
11	0.47	115.34	26	2.27	537.27
12	6.23	121.57	27	15.63	552.90
13	3.39	124.96	28	120.78	673.68
14	9.11	134.07	29	30.81	704.49
15	2.18	136.25	30	34.19	738.68

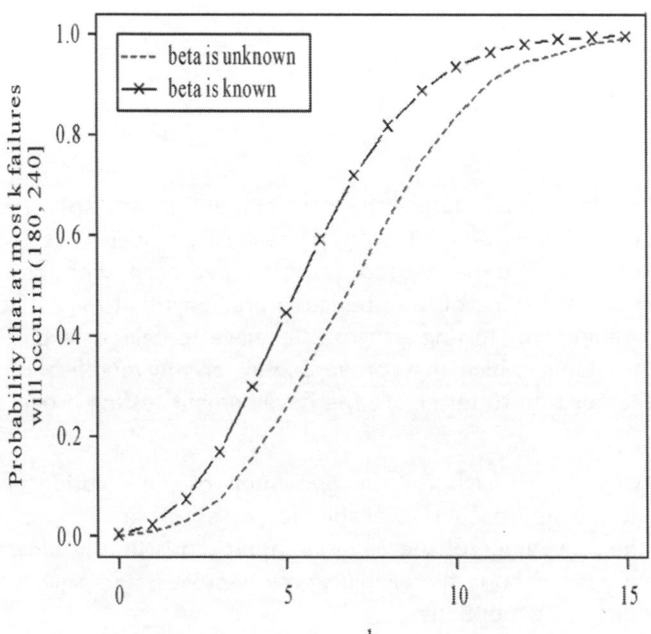

Figure 1. Comparison of the probabilities γ_k that at most k failures will occur in the time interval (180, 240] for the cases of known and unknown β.

3) Since the target value λ_{tv} was not achieved at time $T = 182.21$, we want to know how long it will require in order to attain λ_{tv}. a) When β is known (i.e. $\beta = 0.003962$), let $\gamma = 0.90$, from Equation (14) we obtain $\tau^* = 268.6116h$. In other words, it will take another 268.6116h in order to achieve the desired failure rate. b) When β is unknown, from Equation (15) and Equation (16), we obtain $\tau^* = 770.79\ \mathrm{h}$. In other words, it will take another 770.79 h in order to achieve the desired failure rate when β is unknown.

4) Given $\tau = 900\ \mathrm{h}$, a) when β is known, from Equation (17), the Bayesian Upper Prediction Limit of $\lambda_\tau = \alpha\beta \mathrm{e}^{-\beta\tau}$ with level 0.90 is given by $\lambda_u^{(\beta)}(\tau) = 0.0051$. b) When β is unknown, from Equation (18) and Equation (19), the Bayesian UPL of $\lambda_\tau = \alpha\beta \mathrm{e}^{-\beta\tau}$ with level 0.90 is given by $\lambda_u^{(\beta)}(\tau) = 0.131952$.

5. DISCUSSION

Several prediction problems arise during the development of any software especially when the Goel-Okumoto (1979) software reliability model is used to model the failure process. We have used Bayesian approach with non-informative priors to address some of the prediction problems that may arise during software development testing stage. We have obtained explicit solutions to these problems, which may prove useful for the modification, debugging and for the decision to terminate the development testing process of the software.

The adoption of Bayesian approach for the derivation of the solutions is advantageous in that the approach is available for cases of small sample sizes [11] [12] . Another advantage of the Bayesian approach is that it allows the input of prior information about the reliability growth process and provides full posterior and predictive distributions.

In this paper, we have used non-informative priors to derive the methodologies to address the said prediction problems. However, informative priors can similarly be used in place of non-informative priors. The same procedures presented in this paper can also be applied to other NHPPs such as the delayed S-shaped process and the Cox-Lewis process.

REFERENCES

1. Nuria, T.R. (2011) Stochastic Comparisons and Bayesian Inference in Software Reliability. Ph.D. Thesis, Universidad Carlos III de Madrid, Madrid.
2. Daniel, R.J. and Hoang, P. (2001) On the Maximum Likelihood Estimates for the Goel-Okumoto Software Reliability Model. The American Statistician, 55, 219-222.
3. Meth, M. (1992) Reliability Growth Myths and Methodologies: A Critical View. Proceedings of the Annual Reliability and Maintainability Symposium, New York, 21-23 January 1992, 230-238.
4. Yu, J.-W., Tian, G.-L. and Tang, M.-L. (2007) Predictive Analyses for Nonhomogeneous Poisson Processes with Power Law Using Bayesian Approach. Computational Statistics & Data Analysis, 51, 4254-4268.
5. Kapur, P.K., Pham, H., Gupta, A. and Jha, P.C. (2011) Software Reliability Assessment with OR Applications. Springer Series in Reliability Engineering, Springer-Verlag London Limited, London.
6. Stringfellow, C. and Amschler, A.A. (2002) An Empirical Method for Selecting Software Reliability Growth Models. Empirical Software Engineering, 7, 319-343.
7. Meyfroyt, P.H.A. (2012) Parameter Estimation for Software Reliability Models. Ph.D. Thesis, Universidad Carlos III de Madrid, Madrid.
8. Razeef, M. and Mohsin, N. (2012) Software Reliability Growth Models: Overview and Applications. Journal of Emerging Trends in Computing and Information Sciences, 3, 1309-1320.
9. Xie, M., Goh, T.N. and Ranjan, P. (2002) Some Effective Control Chart Procedures for Reliability Monitoring. Reliability Engineering and System Safety, 77, 143-150.
10. Satya, P., Bandla, S.R. and Kantham, R.R.L. (2011) Assessing Software Reliability Using Inter Failures Time Data. International Journal of Computer Applications, 18, 975-978.
11. Phillips, M.J. (2000) Bootstrap Confidence Regions for the Expected ROCOF of a Repairable System. IEEE Transactions on Reliability, 49, 204-208.

12. Quigley, J. and Walls, L. (2003) Confidence Intervals for Reliability-Growth Models with Small Sample-Sizes. IEEE Transactions on Reliability, 52, 257-262.

Chapter 11

Inferences under a Class of Finite Mixture Distributions Based on Generalized Order Statistics

Abd EL-Baset A. Ahmad[1], Areej M. AL-Zaydi[2]

[1]*Department of Mathematics, Assiut University, Assiut, Egypt*

[2]*Department of Mathematics, Taif University, Taif, KSA*

ABSTRACT

The main purpose of this paper is to obtain estimates of parameters, reliability and hazard rate functions of a heterogeneous population represented by finite mixture of two general components. The doubly Type II censoring of generalized order statistics scheme is used. Maximum likelihood and Bayes methods of estimation are used for this purpose. The two methods of estimation are compared via a Monte Carlo Simulation study

Keywords: Generalized order statistics; Bayes estimation; Heterogeneous population; Monte Carlo Integration; Monte Carlo Simulation

1. INTRODUCTION

Let the random variable (rv) T follow a class including some known lifetime models, its cumulative distribution function (CDF) is given by

$$F(t) = 1 - \exp\left[-\theta\lambda(t)\right], t > 0, (\theta > 0), \tag{1}$$

and its probability density function (PDF) is given by

$$f(t)=\theta\lambda'(t)\exp\left[-\theta\lambda(t)\right],t>0,(\theta>0), \tag{2}$$

where $\lambda'(t)$ is the derivative of $\lambda(t)$ with respect to t and $\lambda(t)\equiv\lambda(t;\alpha)$ is a nonnegative continuous function of t and α may be a vector of parameters, such that $\lambda(t)\to 0$ as $t\to 0^+$ and $\lambda(t)\to\infty$ as $t\to\infty$.

The reliability function (RF) and hazard rate function (HRF) are given, respectively, by

$$R(t)=\exp\left[-\theta\lambda(t)\right], \tag{3}$$

$$H(t)=\theta\lambda'(t), \tag{4}$$

Where

$$H(.)=f(.)/R(.).$$

Bayesian inferences based on finite mixture distribution have been discussed by several authors. Bayesian estimation of the mixing parameter, mean and reliability function of a mixture of two exponential lifetime distributions based on right censored samples considered by [1,2] estimated the survival and hazard functions of a finite mixture of two Gompertz components by using type I and type II censored samples, using the maximum likelihood (ML) and Bayes methods. Based on type I censored samples from a finite mixture of two truncated type I generalized logistic components, [3] computed the Bayes estimates of parameters, reliability and hazard rate functions. [4] considered estimation for the mixed exponential distribution based on record statistics. [5] considered Bayes inference under a finite mixture of two compound Gompertz components model. [6] studied some properties of the mixture of two inverse Weibull distributions and obtained the estimates of the unknown parameters via the EM Algorithm.

[7] introduced the generalized order statistics (gos's). Ordinary order statistics, ordinary record values and sequential order statistics are, among others, special cases of gos's. The gos's have been considered extensively by many authors, among others, they are [8-20].

Mixtures of distributions arise frequently in life testing, reliability, biological and physical sciences. Some of the most important references that discussed different types of mixtures of distributions are a monograph by [21-23].

The PDF, CDF, RF and HRF of a finite mixture of two components of the class under study are given, respectively,

$$f(t) = p_1 f_1(t) + p_2 f_2(t), \tag{5}$$

$$F(t) = p_1 F_1(t) + p_2 F_2(t), \tag{6}$$

$$R(t) = p_1 R_1(t) + p_2 R_2(t), \tag{7}$$

$$H(t) = f(t)/R(t), \tag{8}$$

where, for $j = 1,2$, the mixing proportions p_j are such that $0 \le p_j \le 1, p_1 + p_2 = 1$ and $f_j(t), F_j(t), R_j(t)$ are given from (1), (2), (3) after using θ_j and $\lambda_j(t)$ instead of θ and $\lambda(t)$.

The property of identifiability is an important consideration on estimating the parameters in a mixture of distributions. Also, testing hypothesis, classification of random variables, can be meaning fully discussed only if the class of all finite mixtures is identifiable. Idenifiability of mixtures has been discussed by several authors, including [24-26].

Our aim of this paper is the estimation of the parameters and functions of these parameters of a class of finite mixture distributions based on doubly Type II censoring gos's using ML and Bayes methods. Illustrative example of Gompertz distribution is given and compared with the results obtained by previous researchers.

2. MAXIMUM LIKELIHOOD ESTIMATION

Let

$$T_{s;n,m,k}, T_{s+1;n,m,k}, \cdots, T_{r;n,m,k}, 1 \le s < r \le n, k > 0,$$

be the (r - s) gos's drawn from a mixture of two components of the class (2). Based on this doubly censored sample, the likelihood function can be written [27] as

$$L(\theta|t) = \begin{cases} c_1 \left\{ \prod_{i=s}^{r} \left[R(t_i) \right]^m f(t_i) \right\} \left[R(t_r) \right]^{\gamma_{r+1}} \\ \qquad \times \sum_{\ell=0}^{s-1} \omega_\ell^{(s)} \left[R(t_s) \right]^{(s-\ell-1)(m+1)}, & m \neq -1, \\ c_2 \left[R(t_r) \right]^k \left[\ln R(t_s) \right]^{s-1} \prod_{i=s}^{r} H(t_i), & m = -1, \end{cases} \tag{9}$$

where $t = (t_s, \cdots, t_r)$, $\theta \in \Theta$, Θ is the parameter space, and

$$\left. \begin{array}{l} c_1 = \dfrac{(-1)^{s-1} C_{r-1}}{(m+1)^{s-1} (s-1)!}, \quad c_2 = \dfrac{(-1)^{s-1} k^r}{(s-1)!}, \\ C_{r-1} = \prod_{j=1}^{r} \gamma_j, \quad \gamma_r = k + (n-r)(m+1), \\ \omega_\ell^{(s)} = (-1)^\ell \dbinom{s-1}{\ell}. \end{array} \right\}$$

For definition and various distributional properties of gos's, see [7, 28].

The likelihood function (9) and maximum likelihood estimates (MLE's) can be obtained by using (1) and (5) in two cases, regarding to m value, as follows.

2.1 MLE's When $m \neq -1$

In this case, substituting (1), (5) in (9), the likelihood function takes the form

$$\begin{aligned} L(\theta|t) = c_1 & \left\{ \prod_{i=s}^{r} \left[p_1 R_1(t_i) + p_2 R_2(t_i) \right]^m \right. \\ & \left. \times \left[p_1 f_1(t_i) + p_2 f_2(t_i) \right] \right\} \\ & \times \left[p_1 R_1(t_r) + p_2 R_2(t_r) \right]^{\gamma_{r+1}} \\ & \times \sum_{\ell=0}^{s-1} \omega_\ell^{(s)} \left[p_1 R_1(t_s) + p_2 R_2(t_s) \right]^{(s-\ell-1)(m+1)}. \end{aligned} \tag{10}$$

Take the logarithm of (10), we have

$$\ell(\theta) \equiv \ln L(\theta|\boldsymbol{t}) = \ln c_1 + m\sum_{i=s}^{r}\ln\left[p_1R_1(t_i)+p_2R_2(t_i)\right]$$
$$+\sum_{i=s}^{r}\ln\left[p_1f_1(t_i)+p_2f_2(t_i)\right]+\gamma_{r+1}\ln\left[p_1R_1(t_r)+p_2R_2(t_r)\right]$$
$$+\ln\left\{\sum_{\ell=0}^{s-1}\omega_\ell^{(s)}\left[p_1R_1(t_s)+p_2R_2(t_s)\right]^{(s-\ell-1)(m+1)}\right\}, \tag{11}$$

where $p_1 = p$, $p_2 = 1-p$.

Differentiating (11) with respect to the parameters p, θ_j and α_j (involved in λ) and equating to zero gives the following likelihood equations

$$\left.\begin{aligned}
&\frac{\partial\ell}{\partial p} = m\sum_{i=s}^{r}\vartheta^*(t_i)+\sum_{i=s}^{r}\vartheta(t_i)+\gamma_{r+1}\vartheta^*(t_r)\frac{\sum_{\ell=0}^{s-1}\delta\tau_\ell(t_s)\vartheta^*(t_s)}{\sum_{\ell=0}^{s-1}\tau_\ell(t_s)} \\
&\quad = 0, \\
&\frac{\partial\ell}{\partial\theta_j} = -m\sum_{i=s}^{r}p_j\lambda_j(t_i)\psi_j^*(t_i)+\sum_{i=s}^{r}p_j\xi_j(t_i)\psi_j(t_i) \\
&\qquad -\gamma_{r+1}p_j\lambda_j(t_r)\psi_j^*(t_r) \\
&\qquad -\frac{\sum_{\ell=0}^{s-1}\delta\tau_\ell(t_s)p_j\lambda_j(t_s)\psi_j^*(t_s)}{\sum_{\ell=0}^{s-1}\tau_\ell(t_s)} = 0, \quad j=1,2, \\
&\frac{\partial\ell}{\partial\alpha_j} = -m\sum_{i=s}^{r}p_j\theta_j\frac{\partial\lambda_j(t_i)}{\partial\alpha_j}\psi_j^*(t_i)+\sum_{i=s}^{r}p_j\xi_j^*(t_i)\psi_j(t_i) \\
&\qquad -\gamma_{r+1}p_j\theta_j\frac{\partial\lambda_j(t_r)}{\partial\alpha_j}\psi_j^*(t_r) \\
&\qquad -\frac{\sum_{\ell=0}^{s-1}\delta\tau_\ell(t_s)p_j\theta_j\dfrac{\partial\lambda_j(t_s)}{\partial\alpha_j}\psi_j^*(t_s)}{\sum_{\ell=0}^{s-1}\tau_\ell(t_s)} = 0, \quad j=1,2
\end{aligned}\right\} \tag{12}$$

where, for j = 1,2

$$\left.\begin{array}{l}\xi_j(t_i)=\left[\frac{1}{\theta_j}-\lambda_j(t_i)\right],\ \xi_j^*(t_i)=\left[\frac{1}{\lambda_j'(t_i)}\frac{\partial\lambda_j'(t_i)}{\partial\alpha_j}-\theta_j\frac{\partial\lambda_j(t_i)}{\partial\alpha_j}\right],\ \vartheta(t_i)=\frac{f_1(t_i)-f_2(t_i)}{f(t_i)},\\ \vartheta^*(t_i)=\frac{R_1(t_i)-R_2(t_i)}{R(t_i)},\ \psi_j(t_i)=\frac{f_j(t_i)}{f(t_i)},\ \psi_j^*(t_i)=\frac{R_j(t_i)}{R(t_i)},\ \tau_\ell(t_s)=\omega_\ell^{(s)}R(t_s)^\delta,\delta=(s-\ell-1)(m+1)\end{array}\right\}\tag{13}$$

The solution of the five nonlinear likelihood Equations (12) using numerical method, yields the MLE's $\hat{p},\hat{\theta}_1,\hat{\theta}_2,\hat{\alpha}_1$ and $\hat{\alpha}_2$.

2.2. MLE's When $m=-1$

The likelihood function takes the form

$$L(\theta|t)=c_2\left[p_1R_1(t_r)+p_2R_2(t_r)\right]^k\times\left(\ln\left[p_1R_1(t_s)+p_2R_2(t_s)\right]\right)^{s-1}\times\prod_{i=s}^{r}\frac{\left[p_1f_1(t_i)+p_2f_2(t_i)\right]}{\left[p_1R_1(t_i)+p_2R_2(t_i)\right]}.\tag{14}$$

So, from (14)

$$\begin{aligned}\ell(\theta)\propto k\ln\left[p_1R_1(t_r)+p_2R_2(t_r)\right]&+(s-1)\ln\left(\ln\left[p_1R_1(t_s)+p_2R_2(t_s)\right]\right)\\&+\sum_{i=s}^{r}\left(\ln\left[p_1f_1(t_i)+p_2f_2(t_i)\right]-\ln\left[p_1R_1(t_i)+p_2R_2(t_i)\right]\right).\end{aligned}\tag{15}$$

Differentiating (15) with respect to the parameters p,θ_j and α_j and equating to zero gives the following likelihood equations

$$\left.\begin{array}{l}\frac{\partial\ell}{\partial p}=k\vartheta^*(t_r)+(s-1)\vartheta^*(t_s)\varphi(t_s)+\sum_{i=s}^{r}\left(\vartheta(t_i)-\vartheta^*(t_i)\right)=0,\\ \frac{\partial\ell}{\partial\theta_j}=-kp_j\lambda_j(t_r)\psi_j^*(t_r)-(s-1)p_j\lambda_j(t_s)\psi_j^*(t_s)\varphi(t_s)+\sum_{i=s}^{r}\left(p_j\xi_j(t_i)\psi_j(t_i)+p_j\lambda_j(t_i)\psi_j^*(t_i)\right)\\ \quad=0,\qquad j=1,2,\\ \frac{\partial\ell}{\partial\alpha_j}=-kp_j\theta_j\frac{\partial\lambda_j(t_r)}{\partial\alpha_j}\psi_j^*(t_r)-(s-1)p_j\theta_j\frac{\partial\lambda_j(t_s)}{\partial\alpha_j}\psi_j^*(t_s)\varphi(t_s)+\sum_{i=s}^{r}\left(p_j\xi_j^*(t_i)\psi_j(t_i)+p_j\theta_j\frac{\partial\lambda_j(t_i)}{\partial\alpha_j}\psi_j^*(t_i)\right)\\ \quad=0,\qquad j=1,2\end{array}\right\}\tag{16}$$

where

$$\varphi(t_i) = \frac{1}{\ln R(t_i)}. \tag{17}$$

The solution of the five nonlinear likelihood Equations (16) using numerical method, yields the MLE's $\hat{p}, \hat{\theta}_1, \hat{\theta}_2, \hat{\alpha}_1$ and $\hat{\alpha}_2$.

3. BAYES ESTIMATION

In this section, Bayesian estimation for the parameters of a class of finite mixture distributions is considered under squared error and Linex (Linear-Exponential) loss functions.

We shall use the conjugate prior density, that was suggested by [29], in the following form

$$\begin{aligned} &\pi(\theta;\nu) \propto C(\theta;\nu)\exp[-D(\theta;\nu)], \\ &\theta = (p,\theta_1,\theta_2,\alpha_1,\alpha_2), \nu \in \Omega, \end{aligned} \tag{18}$$

where Ω is the hyperparameter space.

3.1. Bayes Estimates When $m \neq -1$

It follows, from (10) and (18), that the posterior density function is given by

$$\begin{aligned} \pi^*(\theta|t) &= A_1 C(\theta;\nu)\exp[-D(\theta;\nu)] \\ &\times\left\{\prod_{i=s}^{r}[p_1R_1(t_i)+p_2R_2(t_i)]^m[p_1f_1(t_i)+p_2f_2(t_i)]\right\} \\ &\times[p_1R_1(t_r)+p_2R_2(t_r)]^{\gamma_{r+1}} \\ &\times\sum_{\ell=0}^{s-1}\omega_\ell^{(s)}[p_1R_1(t_s)+p_2R_2(t_s)]^{\delta} \end{aligned} \tag{19}$$

Where

$$A_1^{-1} = \int_\theta \pi(\theta;\nu) L(\theta|\boldsymbol{t}) \mathrm{d}\theta. \tag{20}$$

The Bayes estimator of a function, say $u(\theta)$, under squared error and Linex loss functions is given, respectively, by

$$\hat{u}_{BS} = E\left(u(\theta)|\boldsymbol{t}\right) = \int_\theta u(\theta)\pi^*(\theta|\boldsymbol{t})\mathrm{d}\theta, \tag{21}$$

$$\begin{aligned} \hat{u}_{BL} &= \frac{1}{a}\ln\left[E\left(\mathrm{e}^{-au(\theta)}|\boldsymbol{t}\right)\right] \\ &= -\frac{1}{a}\ln\left[\int_\theta \mathrm{e}^{-au(\theta)}\pi^*(\theta|\boldsymbol{t})\mathrm{d}\theta\right], \end{aligned} \tag{22}$$

where the integral is taken over the five dimensional space and $a \neq 0$.

To compute the integral, we can use the Monte Carlo Integration (MCI) method in the form

$$\hat{u}_{BS} = \frac{\sum_{j=1}^{M} u(\theta^j) L(\theta^j|\boldsymbol{t})}{\sum_{j=1}^{M} L(\theta^j|\boldsymbol{t})}, \tag{23}$$

$$\hat{u}_{BL} = -\frac{1}{a}\ln\left[\frac{\sum_{j=1}^{M} \mathrm{e}^{-au(\theta^j)} L(\theta^j|\boldsymbol{t})}{\sum_{j=1}^{M} L(\theta^j|\boldsymbol{t})}\right], \tag{24}$$

where $\theta^j, j=1,2,\cdots,M$ is generated from the PDF $\pi(\theta;v)$, for more details see [30].

Under squared error and Linex loss functions, we can obtain the Bayes estimator of the parameter p, by generating

$$\left(p^j,\theta_1^j,\theta_2^j,\alpha_1^j,\alpha_2^j\right) j=1,2,\cdots,M$$

from the prior (18) and setting $u(\theta)=p$ in (23) and (24). The Bayes estimates of $\theta_1,\theta_2,\alpha_1$ and α_2 can be similarly computed.

3.2. Bayes Estimates When $m=-1$

The posterior density function can be obtained from (14) and (18), as

$$\begin{aligned}\pi^*(\theta|t) &= A_2 C(\theta;v)\exp\left[-D(\theta;v)\right]\\ &\times\left[p_1R_1(t_r)+p_2R_2(t_r)\right]^k\\ &\times\left(\ln\left[p_1R_1(t_s)+p_2R_2(t_s)\right]\right)^{s-1}\\ &\times\prod_{i=s}^{r}\frac{\left[p_1f_1(t_i)+p_2f_2(t_i)\right]}{\left[p_1R_1(t_i)+p_2R_2(t_i)\right]}\end{aligned} \tag{25}$$

Where

$$A_2^{-1}=\int_\theta \pi(\theta;v)L(\theta|t)\,d\theta. \tag{26}$$

Under squared error and Linex loss functions, we can obtain the Bayes estimator of the parameter p, by generating

$$\left(p^j,\theta_1^j,\theta_2^j,\alpha_1^j,\alpha_2^j\right) j=1,2,\cdots,M$$

from the prior (18) and setting $u(\theta)=p$ in (23) and (24). The Bayes estimates of $\theta_1,\theta_2,\alpha_1$ and α_2 can be similarly computed.

4. EXAMPLE

4.1. Gompertz Components

4.1.1. Maximum Likelihood Estimation

Suppose that, for $j = 1,2$ and $t > 0, \theta_j = 1,$

$$\lambda_j(t) = \frac{1}{\alpha_j}\left[e^{\alpha_j t} - 1\right]$$

so

$$\dot{\lambda}_j(t) = e^{\alpha_j t}.$$

In this case, the jth subpopulation is Gompertz distribution with parameter $\alpha_j > 0.$

For $m \neq -1$ by substituting $\lambda_j(t)$ and $\dot{\lambda}_j(t)$ in (12), we have the following nonlinear equations

$$\left.\begin{aligned}
\frac{\partial \ell}{\partial p} &= m\sum_{i=s}^{r} \vartheta^*(t_i) \\
&\quad + \sum_{i=s}^{r} \vartheta(t_i) + \gamma_{r+1}\vartheta^*(t_r) \\
&\quad + \frac{\sum_{\ell=0}^{s-1} \delta\tau_\ell(t_s)\vartheta^*(t_s)}{\sum_{\ell=0}^{s-1} \tau_\ell(t_s)} = 0, \\
\frac{\partial \ell}{\partial \alpha_j} &= -m\sum_{i=s}^{r} p_j\varpi_j(t_i)\psi_j^*(t_i) \\
&\quad + \sum_{i=s}^{r} p_j\xi_j^*(t_i)\psi_j(t_i) \\
&\quad - \gamma_{r+1}p_j\varpi_j(t_r)\psi_j^*(t_r) \\
&\quad - \frac{\sum_{\ell=0}^{s-1} \delta\tau_\ell(t_s)p_j\varpi_j(t_s)\psi_j^*(t_s)}{\sum_{\ell=0}^{s-1} \tau_\ell(t_s)} = 0, \quad j = 1,2
\end{aligned}\right\}, \tag{27}$$

where, for $j = 1, 2$

$$\left.\begin{aligned} \xi_j(t_i) &= 1 - \frac{1}{\alpha_j}\left(e^{\alpha_j t} - 1\right), \\ \xi_j^*(t_i) &= t_i - \varpi_j(t_i), \\ \varpi_j(t_i) &= \frac{1}{\alpha_j^2}\left[1 + \left(\alpha_j t_i - 1\right)e^{\alpha_j t_i}\right] \end{aligned}\right\}, \tag{28}$$

$\hat{p}, \hat{\alpha}_1$ and $\hat{\alpha}_2$ are the solution of the above nonlinear equations.

Also, for $m = -1$ substituting $\lambda_j(t)$ and $\dot{\lambda}_j(t)$ in (13), (16) and (17), we have the following nonlinear equations:

$$\left.\begin{aligned} \frac{\partial \ell}{\partial p} &= k\vartheta^*(t_r) + (s-1)\vartheta^*(t_s)\varphi(t_s) + \sum_{i=s}^{r}\left(\vartheta(t_i) - \vartheta^*(t_i)\right) = 0, \\ \frac{\partial \ell}{\partial \alpha_j} &= -kp_j\varpi_j(t_r)\psi_j^*(t_r) - (s-1)p_j\varpi_j(t_s)\psi_j^*(t_s)\varphi(t_s) + \sum_{i=s}^{r}\left(p_j\xi_j^*(t_i)\psi_j(t_i) + p_j\varpi_j(t_i)\psi_j^*(t_i)\right) = 0, \quad j = 1, 2 \end{aligned}\right\}, \tag{29}$$

$\hat{p}, \hat{\alpha}_1$ and $\hat{\alpha}_2$ are the solution of the above nonlinear equations.

wang#title3_4:spSpecial cases

wang#title3_4:sp1) Upper order statistics

If we put $m = 0$ and $k = 1$ in (10),

$$\left(\gamma_r = n - r + 1\right),$$

the likelihood function takes the form

$$L(\theta|t) = c_1\prod_{i=s}^{r}\left[p_1 f_1(t_i) + p_2 f_2(t_i)\right]\left[p_1 R_1(t_r) + p_2 R_2(t_r)\right]^{n-r} \times \sum_{\ell=0}^{s-1}\omega_\ell^{(s)}\left[p_1 R_1(t_s) + p_2 R_2(t_s)\right]^{s-\ell-1}. \tag{30}$$

Substituting $m=0,k=1$ and $\gamma_r=n-r+1$ in (27), we have the following nonlinear equations

$$\left.\begin{aligned}&\frac{\partial\ell}{\partial p}=\sum_{i=s}^{r}\vartheta(t_i)+(n-r)\vartheta^*(t_r)+\frac{\sum_{\ell=0}^{s-1}\delta\tau_\ell(t_s)\vartheta^*(t_s)}{\sum_{\ell=0}^{s-1}\tau_\ell(t_s)}=0,\\&\frac{\partial\ell}{\partial\alpha_j}=\sum_{i=s}^{r}p_j\xi_j^*(t_i)\psi_j(t_i)-(n-r)p_j\varpi_j(t_r)\psi_j^*(t_r)-\frac{\sum_{\ell=0}^{s-1}\delta\tau_\ell(t_s)p_j\varpi_j(t_s)\psi_j^*(t_s)}{\sum_{\ell=0}^{s-1}\tau_\ell(t_s)}=0,\quad j=1,2\end{aligned}\right\},\tag{31}$$

where $\delta=s-\ell-1$.

The solution of the nonlinear likelihood equations (31) gives the MLE's $\hat{p},\hat{\alpha}_1$ and $\hat{\alpha}_2$.

wang#title3_4:sp2) Upper record values

If we put $k=1$ in (14), $(\gamma_r=1)$ the likelihood function takes the form

$$L(\theta|t)=c_2\left[p_1R_1(t_r)+p_2R_2(t_r)\right]\left(\ln\left[p_1R_1(t_s)+p_2R_2(t_s)\right]\right)^{s-1}\times\prod_{i=s}^{r}\frac{\left[p_1f_1(t_i)+p_2f_2(t_i)\right]}{\left[p_1R_1(t_i)+p_2R_2(t_i)\right]}.\tag{32}$$

Substituting $k=1$ in (29), we have the following nonlinear equations

$$\left.\begin{aligned}&\frac{\partial\ell}{\partial p}=(s-1)\vartheta^*(t_s)\varphi(t_s)+\sum_{i=s}^{r}\vartheta(t_i)-\sum_{i=s}^{r-1}\vartheta^*(t_i)=0,\\&\frac{\partial\ell}{\partial\alpha_j}=\sum_{i=s}^{r}p_j\xi_j^*(t_i)\psi_j(t_i)-(s-1)p_j\varpi_j(t_s)\psi_j^*(t_s)\varphi(t_s)+\sum_{i=s}^{r-1}p_j\varpi_j(t_i)\psi_j^*(t_i)=0,\quad j=1,2\end{aligned}\right\},\tag{33}$$

The solution of the nonlinear likelihood Equations (33) gives the MLE's $\hat{p}, \hat{\alpha}_1$ and $\hat{\alpha}_2$.

4.1.2. Bayes Estimation

Let p, α_1 and α_2 are independent random variables such that $p \sim Beta(b_1, b_2)$ and for $j = 1, 2$, α_j to follow a left truncated exponential density with parameter d_j $(LTE(d_j))$, as used by [2]. A joint prior density function is then given by

$$\pi(\theta; v) \propto p_1^{b_1-1} p_2^{b_2-1} \exp\left[-\sum_{j=1}^{2} \frac{(\alpha_j - 1)}{d_j}\right], \tag{34}$$

where

$$0 < p_1 < 1, \alpha_j > 1, (b_1, b_2, d_1, d_2) > 0$$

and $p_2 = 1 - p_1$.

For $m \neq -1,$ the posterior density function $\pi^*(\theta|t)$ then takes the form

$$\pi^*(\theta|t) \propto p_1^{b_1-1} p_2^{b_2-1} \exp\left[-\sum_{j=1}^{2} \frac{(\alpha_j - 1)}{d_j}\right]$$
$$\times \left\{ \prod_{i=s}^{r} \left[p_1 R_1(t_i) + p_2 R_2(t_i) \right]^m \times \left[p_1 f_1(t_i) + p_2 f_2(t_i) \right] \right\}$$
$$\times \left[p_1 R_1(t_r) + p_2 R_2(t_r) \right]^{\gamma_{r+1}} \sum_{\ell=0}^{s-1} \omega_\ell^{(s)} \left[p_1 R_1(t_s) + p_2 R_2(t_s) \right]^{\delta} \tag{35}$$

For m = −1 the posterior density function $\pi^*(\theta|t)$ then takes the form

$$\pi^*(\theta|t) \propto p_1^{b_1-1} p_2^{b_2-1} \exp\left[-\sum_{j=1}^{2}\frac{(\alpha_j-1)}{d_j}\right]\left[p_1R_1(t_r)+p_2R_2(t_r)\right]^k \times \left(\ln\left[p_1R_1(t_s)+p_2R_2(t_s)\right]\right)^{s-1}\prod_{i=s}^{r}\frac{\left[p_1f_1(t_i)+p_2f_2(t_i)\right]}{\left[p_1R_1(t_i)+p_2R_2(t_i)\right]}. \tag{36}$$

Under squared error and Linex loss functions, we can obtain the Bayes estimator of the parameter p by generating $(p^j, \alpha_1^j, \alpha_2^j)\ j = 1, 2, \cdots, M$ from the prior (34) and setting $u(\theta) = p$ in (23) and (24). The Bayes estimates of α_1 and α_2 can be similarly computed.

wang#title3_4:spSpecial cases

wang#title3_4:sp1) Upper order statistics

If we put $m = 0$ and $k = 1$ in (35), $(\gamma_r = n - r + 1)$, the posterior density function takes the form

$$\pi^*(\theta|t) \propto p_1^{b_1-1} p_2^{b_2-1} \exp\left[-\sum_{j=1}^{2}\frac{(\alpha_j-1)}{d_j}\right]\prod_{i=s}^{r}\left[p_1f_1(t_i)+p_2f_2(t_i)\right]\times\left[p_1R_1(t_r)+p_2R_2(t_r)\right]^{n-r}\sum_{\ell=0}^{s-1}\omega_\ell^{(s)}\left[p_1R_1(t_s)+p_2R_2(t_s)\right]^{\delta}. \tag{37}$$

wang#title3_4:sp2) Upper record values

If we put $k = 1$ in (36), $(\gamma_r = 1)$, the posterior density function takes the form

$$\pi^*(\theta|t) \propto p_1^{b_1-1} p_2^{b_2-1} \exp\left[-\sum_{j=1}^{2}\frac{(\alpha_j-1)}{d_j}\right]\left[p_1R_1(t_r)+p_2R_2(t_r)\right] \times \left(\ln\left[p_1R_1(t_s)+p_2R_2(t_s)\right]\right)^{s-1}\prod_{i=s}^{r}\frac{\left[p_1f_1(t_i)+p_2f_2(t_i)\right]}{\left[p_1R_1(t_i)+p_2R_2(t_i)\right]}. \tag{38}$$

Under squared error and Linex loss functions, we can obtain the Bayes estimator of the parameter p, by generating $(p^j, \alpha_1^j, \alpha_2^j)\ j = 1, 2, \cdots, M$ from the prior (34) and setting $u(\theta)=p$ in (23) and (24). The Bayes estimates of α_1 and α_2 can be similarly computed.

5. SIMULATION STUDY

A comparison between ML and Bayes estimators, under either a squared error or a Linex loss functions, is made using a Monte Carlo simulation study in the two cases upper order statistics and upper record values according to the following steps:

1) For a given values of the prior parameters (b_1, b_2) generate a random value p from the $Beta(b_1, b_2)$ distribution.

2) For a given values of the prior parameters d_j, for $j=1,2,$ generate a random value α_j from the $LTE(d_j)$ distribution.

3) Using the generated values of p, α_1 and α_2, we generate a random sample of size $n=(20,30,50)$ from a mixture of two $Gomp(\alpha_j)$ components, $j=1,2,$ as follows:

- generate two observations u_1, u_2 from $\text{Uniform}(0,1)$.
- if $u_1 \le p$, then

$$t=\frac{1}{\alpha_1}\log\left[1-\alpha_1\log\left(1-u_2\right)\right],$$

Otherwise

$$t=\frac{1}{\alpha_2}\log\left[1-\alpha_2\log\left(1-u_2\right)\right].$$

- repeat above steps n times to get a sample of size n.

4) The sample obtained in Step 3 is ordered.

5) The MLE's of the parameters p, α_1 and α_2 are obtained by solving the nonlinear Equations (31), by using Mathematica 6.

6) Using the generated values of p, α_1 and α_2, we generate upper record values of size $n = (5,8,10)$ from a mixture of two $Gomp(\alpha_j), j = 1,2$ components.

7) The MLE's of the parameters p, α_1 and α_2 are obtained by solving the nonlinear Equations (33), by using Mathematica 6.

8) The Bayes estimates under squared error and Linex loss functions (BES, BEL), of p, α_1 and α_2 are computed, by using MCI forms (23) and (24), respectively.

9) The squared deviations $(\hat{\varepsilon} - \varepsilon)^2$ are computed for different samples and censoring sizes, where ε stands for the parameter and $\hat{\varepsilon}$ its estimate (ML or Bayes).

10) The above Steps (3)-(9) are repeated 1000 times. The averages and the estimated risks (ER) are computed over the 1000 repetitions by averaging the estimates and the squared deviations, respectively.

The computational (our) results were computed by using Mathematica 6.0. In all above cases the prior parameters chosen as $d_2 = 2.3$ $b_1 = 4.5, b_2 = 3.2, d_1 = 1.2,$, which yield the generated values of $p = 0.524855,\ \alpha_1 = 1.47589$ and $\alpha_2 = 2.7255$ (as the true values). The true values of $R(t)$ and $H(t)$ when $t = 0.5$, are computed to be $R(0.5) = 0.414044$ and $H(0.5) =$ 2.80864. The value of the shape parameter a of the Linex loss function is $a = 2$. The averages and the estimated risks (ER) are displayed in Tables 1-4. Figures 1 and 2 represent the estimated risks of the estimates in the case of upper order statistics. Figures 3 and 4 represent the estimated risks of the estimates in the case of upper record values.

Table 1. (Upper order statistics) Averages and Estimated Risks (ER) of the estimates of $\theta=(p,\alpha_1,\alpha_2)$ for different samples and censoring sizes.

n (s,r)	Method	$\bar{\bar{p}}$ $ER(\bar{\bar{p}})$	$\bar{\bar{\alpha}}_1$ $ER(\bar{\bar{\alpha}}_1)$	$\bar{\bar{\alpha}}_2$ $ER(\bar{\bar{\alpha}}_2)$
30 (1, 21)	ML	0.725016 (0.195162)	2.20851 (1.03578)	2.28107 (0.692991)
	BL	0.607509 (0.00766407)	1.065 (0.169001)	1.10095 (2.63928)
	BS	0.630063 (0.0116613)	1.06868 (0.166023)	1.10867 (2.61438)
50 (1, 35)	ML	0.682439 (0.183833)	2.15281 (0.758847)	2.24968 (0.522389)
	BL	0.458447 (0.00464071)	1.04342 (0.187036)	1.05501 (2.79062)
	BS	0.464408 (0.00402117)	1.04364 (0.186845)	1.05637 (2.7861)
20 (1, 20)	ML	0.695272 (0.178062)	1.89954 (0.505669)	2.38476 (0.90054)
	BL	0.562142 (0.00151982)	1.62688 (0.0668917)	1.82389 (0.85426)
	BS	0.589223 (0.00429)	1.97985 (0.360228)	2.83692 (0.411247)
30 (1, 30)	ML	0.671677 (0.1622)	1.79666 (0.353318)	2.45167 (1.18287)
	BL	0.556202 (0.00130063)	1.64213 (0.0834779)	1.83633 (0.838901)
	BS	0.585439 (0.004034)	1.96737 (0.352081)	2.67792 (0.375141)
50 (1, 50)	ML	0.626802 (0.144386)	1.70131 (0.209794)	2.49075 (0.791618)
	BL	0.563246 (0.00174849)	1.651 (0.0792463)	1.8386 (0.821578)
	BS	0.588489 (0.00434187)	1.93554 (0.292391)	2.65172 (0.359827)

n (s,r)	Method	$\bar{\bar{p}}$ $ER(\bar{\bar{p}})$	$\bar{\bar{\alpha}}_1$ $ER(\bar{\bar{\alpha}}_1)$	$\bar{\bar{\alpha}}_2$ $ER(\bar{\bar{\alpha}}_2)$
20 (2, 14)	ML	0.707005 (0.19356)	2.31845 (1.46886)	2.39962 (0.850168)
	BL	0.616284 (0.00864557)	1.09815 (0.14386)	1.13638 (2.52742)
	BS	0.643168 (0.0141694)	1.10888 (0.136649)	1.16379 (2.44375)
30 (2, 21)	ML	0.714324 (0.188562)	2.25684 (1.18964)	2.32837 (0.716555)
	BL	0.632902 (0.011701)	1.022 (0.20618)	1.12869 (2.55009)
	BS	0.683938 (0.0254091)	1.02377 (0.204632)	1.16001 (2.45121)
50 (2, 35)	ML	0.666667 (0.181185)	2.17853 (0.806393)	2.26631 (0.516371)
	BL	0.547597 (0.000574537)	1.0264 (0.20208)	1.06996 (2.74081)
	BS	0.552577 (0.000797349)	1.02767 (0.200941)	1.07139 (2.73608)
20 (1, 14)	ML	0.710467 (0.196735)	2.36646 (1.64159)	2.44219 (0.907496)
	BL	0.601193 (0.005991)	1.10452 (0.138821)	1.1209 (2.57694)
	BS	0.623959 (0.00994848)	1.1156 (0.131404)	1.15205 (2.48091)

Table 2. (Upper order statistics) averages and estimated risks (ER) of the estimates of $R(t)$ and $H(t)$ for different sample and censoring sizes.

$n\ (s,r)$	Method	$\bar{\bar{R}}(t)\ ER(\bar{\bar{R}}(t))$	$\bar{\bar{H}}(t)\ ER(\bar{\bar{H}}(t))$
30 (1, 21)	ML	0.393384 (0.00610191)	3.31504 (2.11435)
	BL	0.515905 (0.0103771)	1.70996 (1.2072)
	BS	0.515913 (0.0103787)	1.71063 (1.20575)
50 (1, 35)	ML	0.398975 (0.00348857)	3.14314 (0.93894)
	BL	0.518317 (0.0108729)	1.68844 (1.25485)
	BS	0.518318 (0.0108731)	1.68851 (1.2547)
20 (1, 20)	ML	0.409763 (0.00335753)	2.962 (0.742865)
	BL	0.394839 (0.00226178)	2.58386 (0.154358)
	BS	0.398332 (0.0020448)	2.95451 (0.319495)
30 (1, 30)	ML	0.413582 (0.00272424)	2.881 (0.51324)
	BL	0.398846 (0.00209693)	2.62951 (0.174472)
	BS	0.401559 (0.00195636)	2.92977 (0.308405)
50 (1, 50)	ML	0.415782 (0.00169979)	2.82688 (0.290495)
	BL	0.402934 (0.00139306)	2.67367 (0.135707)
	BS	0.404751 (0.001321)	2.87568 (0.186433)

$n\ (s,r)$	Method	$\bar{\bar{R}}(t)\ ER(\bar{\bar{R}}(t))$	$\bar{\bar{H}}(t)\ ER(\bar{\bar{H}}(t))$
20 (2, 14)	ML	0.381661 (0.00906794)	3.68514 (7.66423)
	BL	0.511856 (0.00958999)	1.74308 (1.13692)
	BS	0.511923 (0.00960227)	1.7486 (1.12574)
30 (2, 21)	ML	0.388581 (0.00706446)	3.43017 (2.94528)
	BL	0.51866 (0.0109467)	1.68457 (1.26369)
	BS	0.51867 (0.0109488)	1.6854 (1.26185)
50 (2, 35)	ML	0.396371 (0.00370546)	3.18718 (1.07172)
	BL	0.518651 (0.0109425)	1.68541 (1.26163)
	BS	0.518651 (0.0109426)	1.68546 (1.26153)
20 (1, 14)	ML	0.376971 (0.0100604)	3.85478 (12.0168)
	BL	0.511715 (0.00955799)	1.74519 (1.13219)
	BS	0.51177 (0.00956825)	1.7498 (1.12281)

Table 3. (Upper record values) Averages and Estimated Risks (ER) of the estimates of $\theta = (p, \alpha_1, \alpha_2)$ for different sample

$n\ (s,r)$	Method	$\bar{\bar{p}}\ ER(\bar{\bar{p}})$	$\bar{\bar{\alpha}}_1\ ER(\bar{\bar{\alpha}}_1)$	$\bar{\bar{\alpha}}_2\ ER(\bar{\bar{\alpha}}_2)$
8 (1, 7)	ML	0.553391 (0.194865)	2.58457 (1.69481)	2.9875 (2.2188)
	BL	0.553064 (0.00100325)	2.00503 (0.360188)	2.14377 (0.404423)
	BS	0.582772 (0.0035541)	2.56838 (1.36471)	3.76664 (1.4979)
10 (1, 9)	ML	0.737692 (0.195127)	2.12728 (0.627068)	2.69955 (5.2213)
	BL	0.551621 (0.000954124)	1.86627 (0.218747)	2.08316 (0.468689)
	BS	0.583116 (0.00359548)	2.24422 (0.690718)	3.34005 (0.632348)
5 (1, 5)	ML	0.69058 (0.195468)	2.25104 (1.41082)	2.6748 (1.11396)
	BL	0.555479 (0.00102412)	1.72972 (0.116361)	1.91339 (0.699269)
	BS	0.584415 (0.00363311)	2.21588 (0.7209)	3.26564 (0.576103)
8 (1, 8)	ML	0.671123 (0.196826)	2.28674 (1.0187)	2.85737 (4.12699)
	BL	0.538713 (0.000392836)	1.89649 (0.283578)	2.10966 (0.470518)
	BS	0.56659 (0.00190609)	2.4252 (1.13393)	3.41334 (0.749679)
10 (1, 10)	ML	0.789522 (0.194147)	2.00076 (0.478516)	2.67385 (1.92451)
	BL	0.547329 (0.000585634)	1.82498 (0.181567)	2.03864 (0.53141)
	BS	0.575719 (0.00265766)	2.23599 (0.660433)	3.23598 (0.476606)

$n\ (s,r)$	Method	$\bar{\bar{p}}\ ER(\bar{\bar{p}})$	$\bar{\bar{\alpha}}_1\ ER(\bar{\bar{\alpha}}_1)$	$\bar{\bar{\alpha}}_2\ ER(\bar{\bar{\alpha}}_2)$
5 (2, 4)	ML	0.631933 (0.200897)	2.67986 (2.83816)	2.89873 (1.4118)
	BL	0.558404 (0.00128567)	1.74504 (0.108009)	1.98448 (0.579141)
	BS	0.587338 (0.00404486)	2.28454 (0.793391)	3.46485 (0.776266)
8 (2, 7)	ML	0.564415 (0.193903)	2.61393 (2.11598)	3.0953 (5.09421)
	BL	0.545122 (0.000661206)	1.99517 (0.36226)	2.21236 (0.334018)
	BS	0.574543 (0.00269061)	2.59249 (1.47151)	3.70339 (1.32978)
10 (2, 9)	ML	0.751296 (0.19713)	2.14647 (0.632337)	2.93067 (23.1693)
	BL	0.553808 (0.000924086)	1.86508 (0.227293)	2.11734 (0.431788)
	BS	0.580802 (0.00320725)	2.30151 (0.800718)	3.30646 (0.509574)
5 (1, 4)	ML	0.626796 (0.2019)	2.62149 (2.54316)	2.85242 (1.32977)
	BL	0.561015 (0.00133488)	1.74589 (0.104393)	1.92257 (0.667007)
	BS	0.5882	2.27422	3.24856

Table 4. (Upper record values) Averages and Estimated Risks (ER) of the estimates of $R(t)$ and $H(t)$ for different sample and censoring sizes.

n (s,r)	Method	$\bar{\bar{R}}(t)$ $ER\left(\bar{\bar{R}}(t)\right)$	$\bar{\bar{H}}(t)$ $ER\left(\bar{\bar{H}}(t)\right)$
8 (1, 7)	ML	0.338792 (0.012543)	4.31724 (6.94347)
	BL	0.31614 (0.0115378)	2.99744 (0.157374)
	BS	0.323935 (0.0100044)	3.95132 (2.20381)
10 (1, 9)	ML	0.394473 (0.00356277)	3.19236 (1.10688)
	BL	0.350405 (0.00522127)	2.87756 (0.100219)
	BS	0.356443 (0.00444454)	3.48105 (0.768389)
5 (1, 5)	ML	0.379553 (0.0101978)	3.97702 (60.4934)
	BL	0.356758 (0.00535798)	2.66213 (0.0984434)
	BS	0.364529 (0.00435279)	3.41487 (1.22003)
8 (1, 8)	ML	0.372725 (0.0076614)	3.64809 (3.4103)
	BL	0.331679 (0.00880409)	2.93711 (0.172037)
	BS	0.338926 (0.0076069)	3.74408 (1.62913)
10 (1, 10)	ML	0.406745 (0.00339404)	3.01535 (1.01302)
	BL	0.353242 (0.004766)	2.84017 (0.097151)
	BS	0.359484 (0.0040121)	3.38218 (0.568505)

n (s,r)	Method	$\bar{\bar{R}}(t)$ $ER\left(\bar{\bar{R}}(t)\right)$	$\bar{\bar{H}}(t)$ $ER\left(\bar{\bar{H}}(t)\right)$
5 (2, 4)	ML	0.342028 (0.0172756)	5.66515 (237.48)
	BL	0.344783 (0.00674596)	2.68139 (0.0733731)
	BS	0.353593 (0.00543755)	3.59821 (1.31366)
8 (2, 7)	ML	0.335959 (0.0132904)	4.4366 (10.0859)
	BL	0.312673 (0.0125119)	3.05881 (0.196192)
	BS	0.320432 (0.0109407)	4.07878 (2.75144)
10 (2, 9)	ML	0.392196 (0.00425724)	3.26384 (1.69679)
	BL	0.349632 (0.00541351)	2.86293 (0.118538)
	BS	0.355924 (0.00461905)	3.46669 (0.874738)
5 (1, 4)	ML	0.348583 (0.0161086)	4.99658 (64.6011)
	BL	0.352821 (0.00530487)	2.69339 (0.0650445)
	BS	0.360458 (0.00428379)	3.5342 (1.30239)

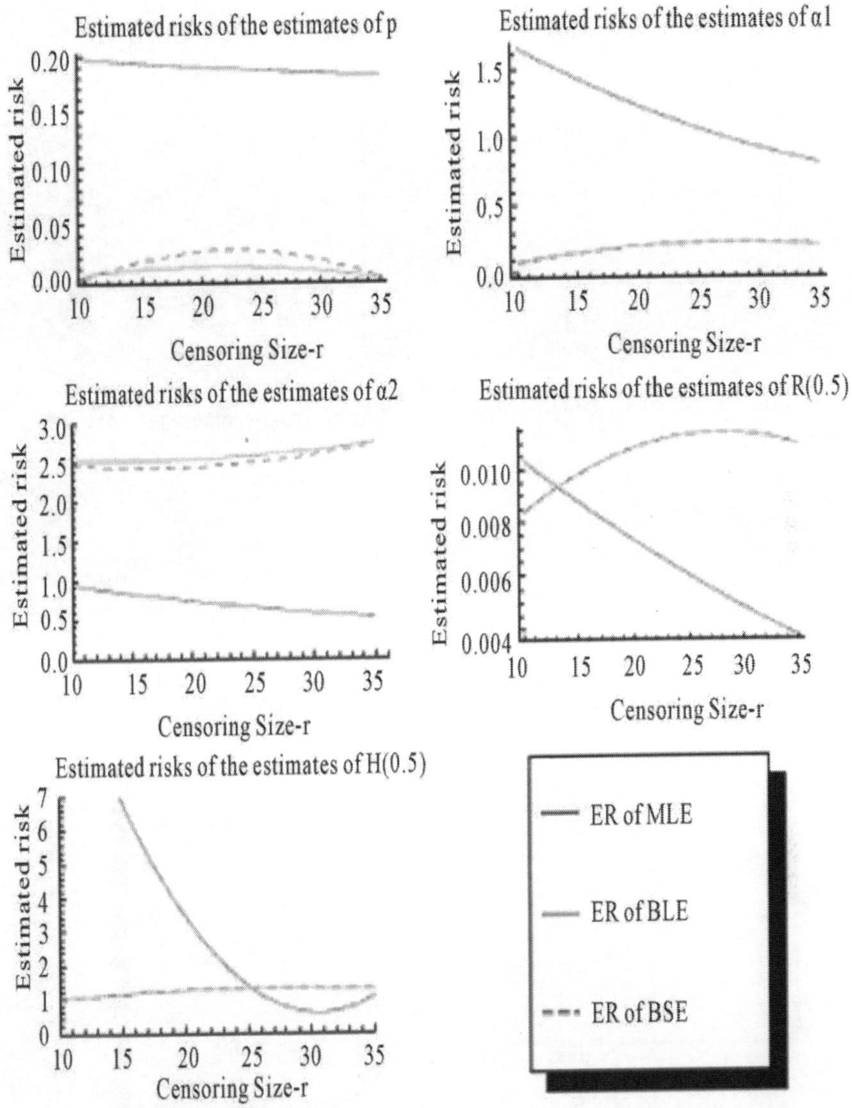

Figure 1. Estimated Risks (ER) of the estimates of $\boldsymbol{\theta} = (p, \alpha_1, \alpha_2)$, $R(t)$ and $H(t)$ based on doubly Type II censored samples (s = 2).

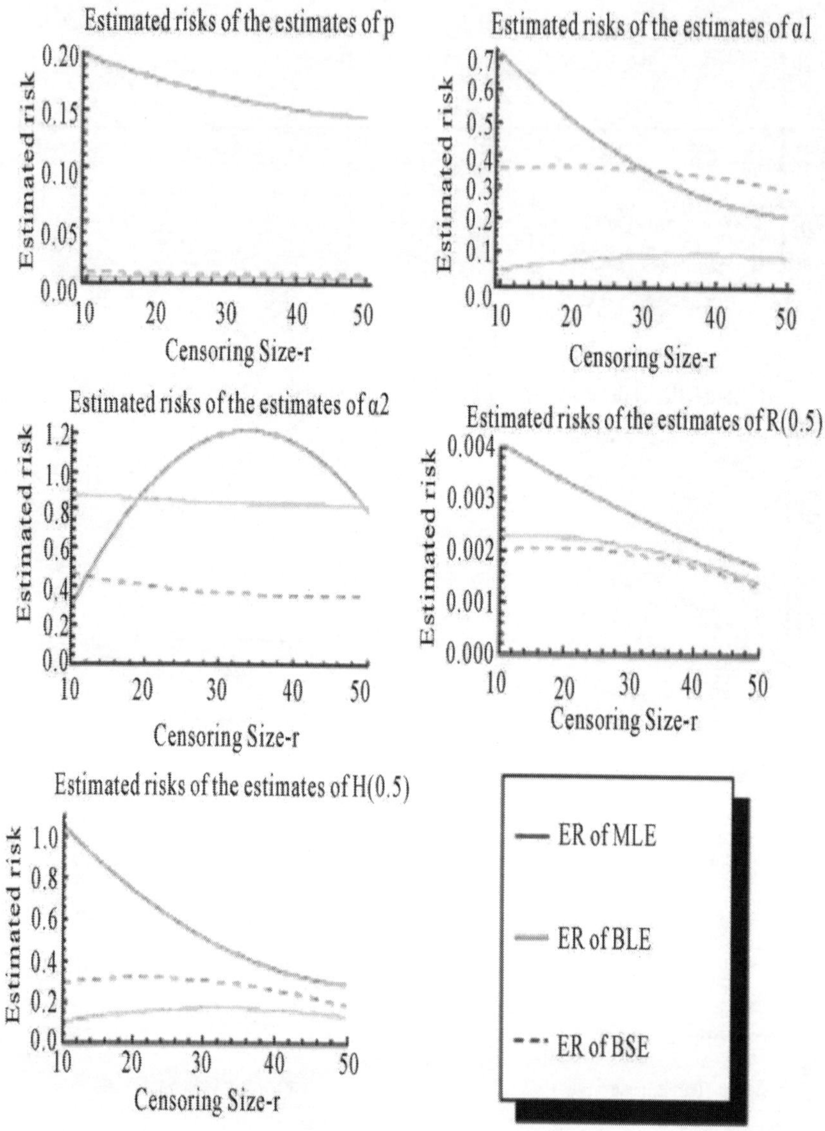

Figure 2. Estimated Risks (ER) of the estimates of $\theta = (p, \alpha_1, \alpha_2)$, $R(t)$ and $H(t)$ complete samples.

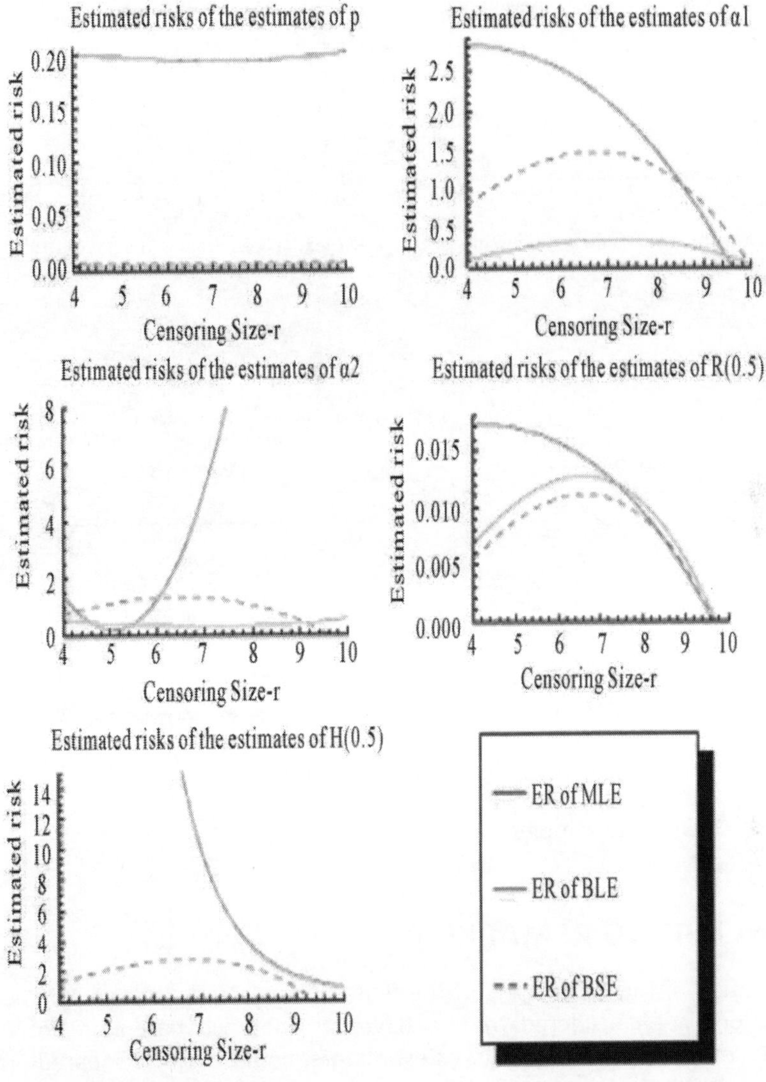

Figure 3. Estimated Risks (ER) of the estimates of $\theta = (p, \alpha_1, \alpha_2)$, $R(t)$ and $H(t)$ based on doubly Type II censored samples (s = 2).

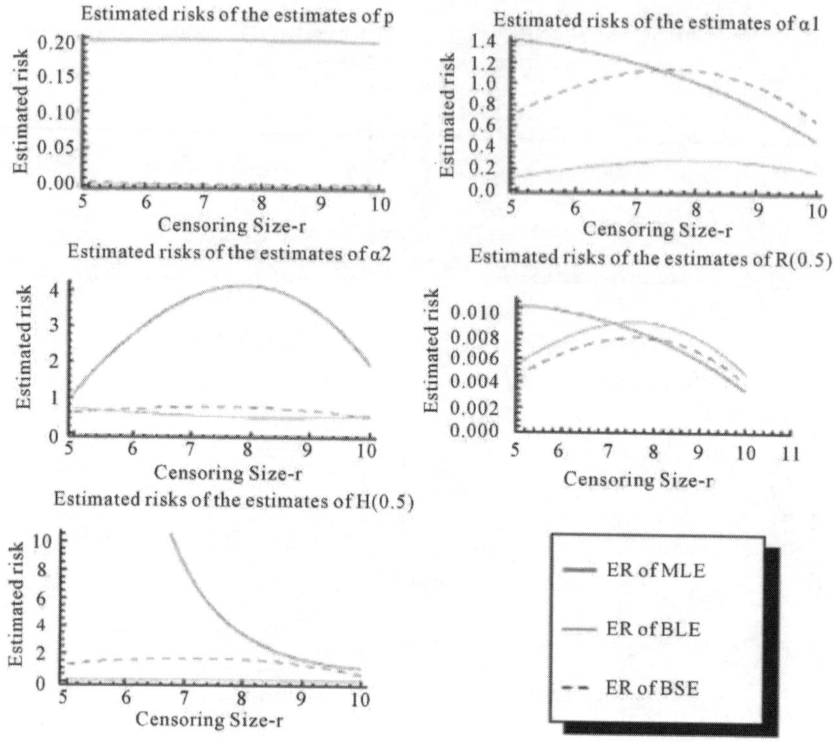

Figure 4. Estimated Risks (ER) of the estimates of $\theta=(p,\alpha_1,\alpha_2)$, $R(t)$ and $H(t)$ complete samples.

6 CONCLUDING REMARKS

1) Estimation of the parameters of the finite mixture model of two Gompertz distributions are considered from a Bayesian approach based on gos's. A compareson between ML and Bayes estimators, under either a squared error loss or a Linex loss, is made by using a Monte Carlo simulation study in both two cases considering order statistics and upper record values cases.

2) From Tables 1 and 2, we see that in most of the considered cases, the ER's of the estimates decrease as n increases. In complete sample case, the Bayes estimates of p, α_1 and HRF under Linex loss function have the smallest ER's as compared with their corresponding estimates under squared error loss function or MLE', while the ER's of the Bayes estimates of α_2 and RF under squared error loss functions are the smallest estimated risks. For censored samples, the Bayes estimates of p under Linex loss function have the smallest

ER's as compared with their corresponding estimates under squared error loss function or MLE's. While, the Bayes estimates (against the proposed prior) of α_1 and HRF under squared error loss function have the smallest ER's as compared with their corresponding estimates. It is observed that MLE's for HRF perform best when sample size n is increased. Also, we note that the MLE's of α_2 and RF have the smallest ER's as compared with Bayes estimates.

3) From Tables 3 and 4, we see that the Bayes estimates (against the proposed prior) of the parameters and HRF under Linex loss function have the smallest ER's as compared with their corresponding estimates under squared error loss function or MLE's. While, the Bayes estimates of α_2 (for complete sample) and RF under squared error loss function have the smallest ER's as compared with both Bayes estimates under Linex loss function or the MLE's. Also, it is observed that MLE's for RF perform best when sample size n is increased.

4) If the mixing proportion p is known, [2] estimated the parameters α_1, α_2, reliability and hazard rate functions based on Types I and II censored samples.

REFERENCES

1. A. S. Papadapoulos and W. J. Padgett, "On Bayes Estimation for Mixtures of Two Exponential-Life-Distributions from Right-Censored Samples," IEEE Transactions on Reliability, Vol. 35, No. 1, 1986, pp. 102-105.

2. E. K. AL-Hussaini, G. R. Al-Dayian and S. A. Adham, "On Finite Mixture of Two-Component Gompertz Lifetime Model," Journal of Statistical Computation and Simulation, Vol. 67, No. 1, 2000, pp. 1-20.

3. E. K. AL-Hussaini and S. F. Ateya, "Bayes Estimations under a Mixture of Truncated Type I Generalized Logistic Components Model," Journal of Applied Statistical Science, Vol. 4 No. 2, 2005, pp. 183-208.

4. Z. F. Jaheen, "On Record Statistics from a Mixture of Two Exponential Distributions," Journal of Statistical Computation and Simulation, Vol. 75, No. 1, 2005, pp. 1- 11.

5. E. K. AL-Hussaini and R. A. AL-Jarallah, "Bayes Inference under a Finite Mixture of Two Compound Gompertz Components Model," Cairo University, 2006.

6. K. S. Sultan, M. A. Ismail and A. S. Al-Moisheer, "Mixture of Two Inverse Weibull Distributions: Properties and Estimation," Computational Statistics & Data Analysis, Vol. 51, No. 11, 2007, pp. 5377-5387.

7. U. Kamps, "A Concept of Generalized Order Statistics," Journal of Statistical Planning and Inference, Vol. 48, No. 1, 1995, pp. 1-23.

8. M. Ahsanullah, "Generalized Order Statistics from Two Parameter Uniform Distribution," Communications in Statistics—Theory and Methods, Vol. 25, No. 10, 1996, pp. 2311-2318.

9. U. Kamps and U. Gather, "Characteristic Property of Generalized Order Statistics for Exponential Distributions," Applied Mathematics (Warsaw), Vol. 24, No. 4, 1997, pp. 383-391.

10. E. Cramer and U. Kamps, "Relations for Expectations of Functions of Generalized Order Statistics," Journal of Statistical Planning and Inference, Vol. 89, No. 1-2, 2000, pp. 79-89.

11. M. Habibullah and M. Ahsanullah, "Estimation of Parameters of a Pareto Distribution by Generalized Order Statistics," Communications in Statistics—Theory and Methods, Vol. 29, No. 7, 2000, pp. 1597-1609.

12. E. K. AL-Hussaini and A. A. Ahmad, "On Bayesian Predictive Distributions of Generalized Order Statistics," Metrika, Vol. 57, No. 2, 2003, pp. 165-176.

13. E. K. AL-Hussaini, "Generalized Order Statistics: Prospective and Applications," Journal of Applied Statistical Science, Vol. 13, No. 1, 2004, pp. 59-85.

14. Z. F. Jaheen, "On Bayesian Prediction of Generalized Order Statistics," Journal of Applied Statistical Science, Vol. 1, No. 3, 2002, pp.191-204.

15. Z. F. Jaheen, "Estimation Based on Generalized Order Statistics from the Burr Model," Communications in Statistics—Theory and Methods, Vol. 34, No. 4, 2005, pp. 785-794.

16. A. A. Ahmad, "Relations for Single and Product Moments of Generalized Order Statistics from Doubly Truncated Burr Type XII Distribution," Journal of the Egyptian Mathematical Society, Vol. 15, No. , 2007, pp. 117- 128.

17. A. A. Ahmad, "Single and product Moments of Generalized order Statistics from Linear Exponential Distribution," Communications in Statistics—Theory and Methods, Vol. 37, No. 8, 2008, pp. 1162- 1172.

18. S. F. Ateya and A. A. Ahmad, "Inferences Based on Generalized Order Statistics under Truncated Type I generalized Logistic Distribution," Statistics, Vol. 45, No. 4, 2011, pp. 389-402.

19. T. A. Abu-Shal and A. M. AL-Zaydi, "Estimation Based on Generalized Order Statistics from a Mixture of Two Rayleigh Distributions,"

International Journal of Statistics and Probability, Vol. 1, No. 2, 2012, pp. 79-90.

20. T. A. Abu-Shal and A. M. AL-Zaydi, "Prediction Based on Generalized Order Statistics from a Mixture of Rayleigh Distributions Using MCMC Algorithm," Open Journal of Statistics, Vol. 2, No. 3, 2012, pp. 356-367.

21. B. S. Everitt and D. J. Hand, "Finite Mixture Distributions," University Press, Cambridge, 1981.

22. D. M. Titterington, A. F. M. Smith and U. E. Makov, "Statistical Analysis of Finite Mixture Distributions," John Wiley and Sons, New York, 1985.

23. G. J. McLachlan and K. E. Basford, "Mixture Models: Inferences and Applications to Clustering," Marcel Dekker, New York, 1988.

24. H. Teicher, "Identifiability of Finite Mixtures," The Annals of Mathematical Statistics, Vol. 34, No. 4, 1963, pp. 1265-1269.

25. E. K. AL-Hussaini and K. E. Ahmad, "On the Identifiability of Finite Mixtures of Distributions," Transactions on Information Theory, Vol. 27, No. 5, 1981, pp. 664- 668.

26. K. E. Ahmad, "Identifiability of Finite Mixtures Using a New Transform," The Annals of Mathematical Statistics, Vol. 40, No. 2, 1988, pp. 261-265.

27. A. A. Ahmad, "On Bayesian Interval Prediction of Future Generalized Order Statistics Using Doubly Censoring," Statistics, Vol. 45, No. 5, 2011, pp. 413- 425.

28. U. Kamps, "A Concept of Generalized Order Statistics," Journal of Statistical Planning and Inference, Vol. 48, No. 1, 1995, pp. 1-23.

29. E. K. AL-Hussaini, "Prediction, Observables from a General Class of Distributions," Journal of Statistical Planning and Inference, Vol. 79, No. 1, 1999, pp. 79-91.

30. S. J. Press, "Subjective and Objective Bayesian Statistics: Principles, Models and Applications," Wiley, New York, 2003.

Chapter 12

Advanced Statistical Tools for Enhanced Quality Digital Imaging with Realistic Capture Models

Aleksandra Pižurica[1*], Javier Portilla[2], Keigo Hirakawa[3] and Karen Egiazarian[4]

[1]*Department of Telecommunications and Information Processing, Ghent University, Gent 9000, Belgium.*

[2]*Imaging and Vision Department, Institute of Optics (CSIC), Madrid 28006, Spain.*

[3]*Electrical and Computer Engineering, University of Dayton, Dayton, OH 45469, USA.*

[4]*Department of Signal Processing, Tampere University of Technology, Tampere 33101, Finland.*

EDITORIAL

Getting closer to reality in modeling image capture devices is crucial for the improvement of image quality beyond the limits of image restoration algorithms as we know them today. This calls for more accurate statistical modeling of distortions and noise coming from real capture devices (Poisson noise, internal non-linearities, space variant point spread functions due to nonideal optics, chromatic aberrations, etc.). While these effects are often not considered in the restoration algorithms, their impact on the resulting image quality is huge in practice. For example, different nonlinearities (both intrinsic to the imaging device and induced ones, e.g., to make the noise signal independent) can invalidate typically assumed noise models and can also devastate deblurring. Joint modeling of digital and nondigital components (like optics and sensors) or various sources of image distortions (such as color filter array, blur, and noise) will likely yield improvements over the traditional approach to treat them separately. Rapid progress in digital camera technology makes a huge impact on computer vision, surveillance and security systems, production of portable electronic devices, such as smart phones. Physical limits of the sensors are likely to impose trade-offs on picture quality (e.g., in terms of achievable resolution versus noise considerations) that can be dealt with only by smart and device-aware signal processing. Moreover, new challenges arise from new imaging technologies, like in three-dimensional (3D) digital cameras [1, 2], and new acquisition modalities, such as compressive sensing [3, 4, 5, 6, 7].

One of the central topics to this special issue is realistic modeling of noise in digital cameras. With ongoing miniaturization, sensor elements of the camera are becoming increasingly sensitive to noise. In state-of-the-art image sensors, the pixel size is approaching 1 μ m. By shrinking the pixel size towards the wavelengths of light that the camera captures, the amount of light received will be further decreased by technical barriers (diffraction effects) [8]. Hence, increasing further the sensor resolution by itself will not necessarily lead to actual gains in image quality. Also, recent improvements in sensor sensitivity allow cameras to operate in very low lighting conditions, but this boosts noise in the acquired images. The negative effects of noise can be largely suppressed by post-processing algorithms, but these require precise knowledge of the noise characteristics to achieve optimal performance. Due to the mismatch between the actual noise characteristics and typically assumed, simplified noise models, the noise level is typically over- or underestimated, affecting adversely the performance of noise reduction.

Three papers in this special issue [9, 10, 11] address the problem of camera noise modeling and different aspects of noise reduction, ranging from post-processing steps for hardware solutions to purely signal processing post-production approaches. Available technologies for reducing noise in hardware include pixel binning [12, 13], where electrical charges of neighboring pixels are combined to form a superpixel. The benefits are faster readout speeds and improved signal-to-noise ratios albeit at the price of spatial resolution loss. Binning in color images is particularly challenging due to the presence of color filter array (CFA). To maintain color fidelity, the binning process combines neighboring pixels with the same color filter, such that the resulting superpixels form again a CFA, typically the Bayer pattern. The full-color image that is recovered from these subsampled sensor data by means of classical demosaicking is not only of lower resolution (compared to the case without binning), but suffers also from characteristic pixelization artifacts. Jin and Hirakawa provide in [9] a comprehensive characterization of pixel binning for color image sensors and propose post-capture signal processing steps to eliminate the binning artifacts. By a rigorous analysis, they prove that pixelization arises due to the mismatch between binning and demosaicking and that it cannot be eliminated simply by increasing the sensor resolution, but rather by a carefully designed binning-aware demosaicking approach.

The contribution by Aiazzi et al. [10] revisits optoelectronic noise model in view of recent advances in the technology of modern CCD color cameras. The authors present an original and robust method for estimating the parameters of a mixed photon and electronic noise from a single image. The main idea behind the method is to decouple the power of the signal-dependent photon noise from that of the signal-independent electronic noise in the space defined by sample mean versus variance, such that noise parameters can be estimated elegantly, by multivariate linear regression. The paper gives a comprehensive discussion on the range of validity of the optoelectronic noise model, clipping considerations in actual cameras and nonlinearities that dominate the camera response function (CRF), which gives new interesting insights with respect to related approaches like [14, 15]. Motivated by this analysis, the authors advocate noise estimation on the CRF-corrected data, i.e., after compensating for the nonlinearities introduced by the electronic chain, showing that in this case potential problems with the range of validity of the model are avoided.

Goossens et al. [11] take a different approach and present a detailed model of the camera noise, related to [16], which not only models a mixture of photon and electronic noise but also takes explicitly into account different aspects, such as amplification (ISO sensitivity of the camera), fixed pattern noise, clipping and quantization due to A/D conversion, and different post-processing operations through the CRF. This noise model is then applied to the reconstruction of high dynamic range images from a small set of low dynamic range acquisitions of a static scene. An original contribution of this work is also a novel Bayesian formulation of the weighting function for combining low dynamic range components into a high dynamic range image.

Another central and rapidly evolving field addressed in this special issue is image restoration. No imaging device has a perfect one-to-one correspondence of spatial directions in the scene (those ideally producing a single projection dot on the output image) and spatial locations in the resulting image. Instead, each of these incoming directions gives rise to a point spread function (PSF, a small 'cloud'), whose shape and size depends (besides the incoming light wavelengths, in refraction-based optical devices) on both the image location from where the ray starts and on the imaging device itself (e.g., on its optical response, for devices capturing visible light). Other practically important source of blur is relative movement between the camera and objects in the scene. Such movements produce dynamically evolving images, which are temporally integrated by the sensor during the exposure time interval. Whereas all of these image degradations have been traditionally modeled in the image processing literature as simple convolutions, this is only a crude first approximation. In fact, the blur phenomenon is better modeled as a 3D field of PSFs depending on the relative coordinates (and movement) of the scene with respect to the camera.

This type of degradation, to a certain extent, is unavoidable in real imaging devices, and, coupled with the also omnipresent imaging noise, poses a formidable challenge for image processing scientists and engineers. The challenge at stake consists of correctly modeling and compensating for the various degradation sources, whereas keeping computational complexity of the restoration reasonably low.

In this special issue, we include four stimulating papers addressing different problems in image restoration. Santiago et al. [17] deal with the problem of boundary handling in image restoration. Despite its practical impact, this problem is often ignored in the restoration field, as evidenced by artificial boundary conditions such as circular or mirror-like boundary conditions used in typical simulation studies. Among the published works really addressing this problem, the model-based ones usually consider an output image with a smaller spatial support than the input image. Whereas this is a conceptually correct approach, it usually leads to highly complex (and, thus, somehow unpractical in most cases) restoration algorithms. Santiago et al. [17], instead, propose here a novel practical approach based on a multilayer perceptron (the classical neural network), which learns the degradation model. More regular results are produced in the total variation sense, without *a priori* constraints on the structure of the image or the boundary conditions. Compared to the previous methods, they achieve robust and more visually plausible reconstructions of the image on the boundary affected areas.

Seo and Milanfar [18] consider the problem of combining a short-exposure image obtained with flash and a long-exposure image taken without flash. The second is generally blurry due to the camera shake during exposure and also noisy due to poor illumination. The goal is to obtain the best of the two of them, i.e., the natural appearance (soft shadows, warm colors, etc.) of the one without flash, and the good definition and low noise of the one with flash. They propose a generalization of the guided filter approach of He et al. [19]. The process involves adjusting an unsupervised prediction model, which has both nonlinear components (reaction-diffusion driven) and a locally linear adaptive dependency. This novel approach results in a significant visual improvement with respect to previous methods.

Miraut et al. present two papers [20, 21] dealing with spatially variant PSF characterization and its corresponding image restoration. In both papers, the authors account for the fact that the PSFs affecting images are not spatially uniform for real imaging devices, but they rather change from one location of the acquired image to another one. In cases specified by the authors, when the PSF does not experience an abrupt change, a smooth PSF field may be used for globally characterizing the blur. In the first paper [20], Miraut and Portilla propose a novel model for representing smooth PSF fields, and for dealing with them efficiently, by means of deformable filtering techniques. A practical, but still model-based and general, methodology is developed on how to perform image restoration on a given image affected by a known, given, smooth PSF field. In the second paper [21], Miraut et al. address the problem of estimating the PSF field from a single observed image. This is done for the especially favorable case of star field images, where the objects are dot-like. Here, the assumption of a smooth PSF field is correct, and in addition, local PSFs are easy to estimate from the stars (almost ideal dot-like objects). Some real examples using ground-based telescope images show the ability of the proposed algorithm to effectively characterize the PSF field.

We close this special issue by two papers [22, 23] devoted to the topic of rapidly emerging sparse reconstruction and compressed sensing approaches. Puy et al. present in [22] a 'spread spectrum' compressed sensing strategy. The main idea behind this approach is to apply a wide bandwidth pre-modulation (i.e., chirp modulation) to the signal of interest before projecting it onto randomly selected vectors of an orthonormal basis. Authors consider the case of Fourier imaging, where pre-modulation with a wide-band signal amounts to a convolution in the Fourier domain, spreading the power spectrum of the original signal, while preserving its norm. Consequently, coherence between the sparsity and the sensing bases is drastically reduced, leading to enhanced reconstruction quality. Enhancement in the reconstruction quality stemming from this approach is demonstrated in two concrete applications: radio interferometric imaging and magnetic resonance imaging.

Finally, Dadkhahi et al. [23] offer a different approach to sparse representations: a reprojection of a signal represented in one basis onto another basis. In particular, the authors concentrate on constructing the sparse representation of piecewise smooth signals using the discrete cosine transform (DCT) by deriving the inverse polynomial reconstruction method for the DCT expansion. They show that this approach enables recovering piecewise smooth signals from a relatively small number of DCT coefficients with high accuracy. The paper demonstrates benefits of this framework in applications of signal and image denoising and approximation.

REFERENCES

1. Aguirre A, Batur A, Hewes G, Pekkucuksen I, Venkatraman N, Ware F, Zhang B: *Embedded stereoscopic 3D camera system for mobile platforms. IEEE International Conference on Acoustics Speech and Signal Processing ICASSP, Kyoto, 25-30 March 2012*. Piscataway: IEEE; 2012.

2. Zhang C, Xu J, Xi N, Yunyi J, Li W: *Development of an omni-directional 3D camera for robot navigation. IEEE/ASME International Conference on Advanced Intelligent Mechatronics (AIM), Kachsiung, 11-14 July 2012*. Piscataway: IEEE; 2012.

3. Candès E, Romberg J, Tao T: Robust uncertainty principles: exact signal reconstruction from highly incomplete frequency information. *IEEE Trans. Inf. Theory* 2006, 52(2):489-509.

4. Chen H, Xi N, Song B, Chen L, Zhao J, Wai Chiu Lai K, Yang R: Infrared camera using a single nano-photodetector. *IEEE Sensors J* 2013, 13(3):949-958.

5. Coluccia G, Kamden-Kuiteng S, Abrardo A, Barni M, Magli E: Progressive compressed sensing and reconstruction of multidimensional signals using hybrid transform/prediction sparsity model. *IEEE J. Emerg. Selected Top. Circuits Syst* 2012, 2(3):340-352.

6. Donoho D: Compressed sensing. *IEEE Trans. Inf. Theory* 2006, 52(4):1289-1306.

7. Sankaranarayanan A, Studer C, Baraniuk R: *CS-MUVI: Video compressive sensing for spatial-multiplexing cameras. IEEE International Conference on Computational Photography, Seattle, 28-29 April 2012*. Piscataway: IEEE; 2012.

8. Nishiwaki S, Nakamura T, Hiramoto M, Fujii T, Suzuki M: Efficient colour splitters for high-pixel-density image sensors. *Nat. Photonics* 2013, 7(3):240-246.

9. Jin X, Hirakawa K: Analysis and processing of pixel binning for color image sensor. *EURASIP J. Adv. Signal Process* 2012, 2012: 125.

10. Aiazzi B, Alparone L, Baronti S, Selva M, Stefani L: Unsupervised estimation of signal-dependent CCD camera noise. *EURASIP J. Adv. Signal Process* 2012, 2012: 231.

11. Goossens B, Luong H, Aelterman J, Pižurica A, Philips W: Realistic camera noise modeling with application to improved HDR synthesis. *EURASIP J. Adv. Signal Process* 2012, 2012: 171.

12. Westra A, Heemskerk J, Korevaar M, Theuwissen A, Kreuger R, Ligtvoet K, Beekman F: On-chip pixel binning in photon-counting EMCCD-based gamma camera: a powerful tool for noise reduction. *IEEE Trans. Nuclear Sci* 2009, 56(5):2559-2565.

13. Zhou Z, Pain B, Fossum E: Frame-transfer CMOS active pixel sensor with pixel binning. *IEEE Trans. Electron. Dev* 1997, 44(10):1764-1768.

14. Foi A, Trimeche M, Katkovnik V, Egiazarian K: Practical Poissonian-Gaussian noise modeling and fitting for single-image raw data. *IEEE Trans. Image Process* 2008, 17(10):1737-1754.

15. Liu C, Szeliski R, Kang S, Zitnick C, Freeman W: Automatic estimation and removal of noise from a single image. *IEEE Trans. Patt. Anal. Mach. Intell* 2008, 30(2):299-314.

16. Hasinoff S, Durand F, Freeman W: *Noise-optimal capture for high dynamic range photography. IEEE 23rd Conference on Computer Vision and Pattern Recognition (CVPR), San Francisco, 13-18 June 2010.* Piscataway: IEEE; 2010.

17. Santiago M, Cisneros G, Bernués E: Boundary reconstruction process of a TV-based neural net without prior conditions.*EURASIP J. Adv. Signal Process* 2011, 2011: 115.

18. Seo HJ, Milanfar P: Robust flash denoising/deblurring by iterative guided filtering. *EURASIP J. Adv. Signal Process* 2012, 2012: 3.

19. He K, Sun J, Tang X: Guided image filtering. In *Computer Vision - ECCV 2010, ed. by K Daniilidis, P Maragos, N Paragios. Proceedings of the 11th European Conference on Computer Vision, Heraklion, Crete, 5-11 September 2010. Lecture Notes in Computer Science, vol. 6311.* Heidelberg: Springer; 2010:1-14.

20. Miraut D, Portilla J: Efficient shift-variant image restoration using deformable filtering (Part I). *EURASIP J. Adv. Signal Process*2012, 2012: 100.

21. Miraut D, Ballé J, Portilla J: Efficient shift-variant image restoration using deformable filtering (Part II): PSF field estimation.*EURASIP J. Adv. Signal Process* 2012, 2012: 193.

22. Puy G, Vandergheynst P, Gribonval R, Wiaux Y: Universal and efficient compressed sensing by spread spectrum and application to realistic fourier imaging techniques, EURASIP Journal on Advances in Signal Processing. *EURASIP J. Adv. Signal Process*2012, 2012: 6.

23. Dadkhahi H, Gotchev A, Egiazarian K: Inverse polynomial reconstruction method in DCT domain. *EURASIP J. Adv. Signal Process* 2012, 2012: 133.

Chapter 13

Applying Statistical Process Control in Small Sized Evolutionary Projects: Results and Lessons Learned in The Implementation of CMMI-DEV Maturity Level 5 in Synapsis Brazil

Carlos Simões[1] and Mariano Montoni[2]*

[1] Synapsis Brasil, S.A., Rua São Pedro n 181, CEP 24.020-054 Centro, Niterói, RJ, Brasil.

[2] ProMove – Business Intelligence Solutions, Rua da Assembleia, No 10, Sala 2805, CEP 22.011-000 Centro, Rio de Janeiro, RJ, Brasil.

ABSTRACT

This paper presents the roadmap adopted by Synapsis Brazil to implement the process areas of CMMI-DEV Maturity Level 5 in small sized evolutionary projects. We also discuss the results of applying statistical process control in more than 20 projects, and the lessons learned in conducting successfully a SCAMPI A assessment of CMMI-DEV Maturity Level 5.

BACKGROUND

Process maturity models like CMMI-DEV and Brazilian model MPS.BR are adopted by organizations as a reference for software process improvement. Many benefits are observed by organizations that adopted such models, for instance, increase in software development team productivity, reduction of projects schedule and costs, and increase in software product quality (Travassos and Kalinowski 2012).

Nonetheless, CMMI-DEV and MPS.BR assessments concentrate on the lower maturity levels, making more difficult to obtain benefits with implementation of higher maturity level best practices. According to (SOFTEX 2013), less than 2% (8 out of 440) of all MPS.BR assessments conducted in Brazil were on the higher maturity level (level A), and 88% of the assessments concentrated on the first two maturity levels (G and F). According to (SEI 2012), only 6% (11 out of 174) of the official valid assessments of SCAMPI A of CMMI-DEV model in Brazil were on the highest maturity level (level 5), while 50% of these assessments concentrated on the lowest two maturity levels (1 and 2).

Many factors difficult the successful execution of software process improvement initiatives. According to (Montoni and Rocha 2011), many of these factors are related to software organization members (for instance, easy to accept changes and higher management support), and also to factors related to consultancy organization members (for instance, knowledge, experience and skills of members of the consultancy) and factors related to organizational strategy and policies (for instance, inexistence of conflict interest in the software process implementation initiative), among others. Therefore, it is likely that many of these factors may also difficult the conduction of software process initiatives in higher maturity levels, like CMMI level 5 and MPS level A.

This paper presents the roadmap adopted by Synapsis Brazil to overcome critical success barriers of the implementation of process areas of level 5 of CMMI-DEV in small sized evolutionary projects. The paper also discusses the results of the application of statistical process control in more than 20 projects, and the lessons learned in the successful conduction of a SCAMPI A assessment of CMMI-DEV maturity level 5 after only 12 months of process implementation with the support of ProMove, a consultancy specialized in software process improvement.

In the CMMI maturity levels 2 and 3 and MPS.BR maturity levels G to C the control of the software process is reactive. In higher maturity levels (CMMI levels 4 and 5, for instance) software process control shall be proactive, based on its performance (Luo et al. 2008; Florac et al. 2000). In this context, traditional software measurement methods and techniques such as comparing planned versus actual data are not sufficient to predict process performance.

One of the mayor difficulties to achieve higher maturity levels in software development is the lack of adequacy of the collected measures and measurement database available in software organizations to apply statistical process control, and the difficulty to access project data (Kitchenham et al. 2006). Therefore, organizations seeking to achieve high maturity in their process shall measure software process efficiently since the early stages of a process improvement program.

Some organizations get frustrated with the assessment of process performance by using statistical process control mainly because they start using these powerful techniques to measure large processes, composed of a great variety of subprocesses (Florac et al. 2000). Statistical process control do not need to be applied in all processes or process components that compose the organization software development process, but it shall be applied only to critical process and not all statistical process control techniques are applicable to all software processes (Sargut and Demirors 2006).

Statistical process control focus in controlling key processes and process components. Moreover, in order to be effective certain requirements must be satisfied, such as reasonable amount of observations, existence of controlled processes and performance objectives aligned to business strategy, definition of measures associated to activities that produce tangible products, and product quality characteristics defined by the clients should be used to select and define measures. Moreover, a process or process component should be easy to control, i.e., should be defined appropriately and of short duration (Sargut and Demirors 2006; Weller et al. 2008).

CASE DESCRIPTION

Synapsis Brazil is one of the most relevant Latin America IT services and technology integration organization, with a wide portfolio of IT services that include IT outsourcing, cloud computing services, IT consultancy and SAP, software factory, technical and business solutions, smart grid solutions and development/implementation of customized solutions. Synapsis Brazil is also leader in specific segments as utilities, oil and gas, and Brazilian govern sectors. The company is also aggressively expanding in other vertical industries with significant presence in Argentina, Brazil, Chile, Colombia and Peru.

In order to demonstrate excellence in software development, Synapsis Brazil invested highly in the last years in the improvement of its processes. The organization was successfully assessed in CMMI maturity levels 2 and 3 in August 2006 and July 2009, respectively. In December 2011, Synapsis Brazil higher management established as a process objective to achieve in December 2012, the official assessment of CMMI maturity level 5. This objective was considered ambitious since the average of the organizations take 2 years to move from CMMI level 3 to 5 (SEI – Software Engineering Institute 2012) (double of the time established by Synapsis Brazil higher management). Even though the objective was ambitious, Synapsis Brazil conducted successfully an official SCAMPI A appraisal on CMMI maturity level 5 in December 2012, after 12 month of process improvement implementation.

One of the factors that most contributed to the success of the initiative was the higher management support. Since the beginning of the implementation, Synapsis Brazil higher management dedicated sufficient human and financial resources to successfully implement CMMI level 5 process areas. Moreover, they participated actively of all meetings to analyze the situation of organization process performance. The support of higher management was also essential in the establishment of quality and process performance objectives. After many meeting with the process group, it was established as performance objective that the project real productivity should be equal or higher than 90% of the expected productivity, whilst the quality objective were that the client acceptance test approval should be equal or higher than 95%. Table 1 presents the evolution of Synapsis Brazil processes to achieve these objectives.

Table 1. Evolution of Synapsis Brazil processes to high maturity

Baseline number	Publication date	Project name	Process improvement objective	Results achieved	CMMI level
14 - 19	May to December 2011	Input Carga de Vendas Input Outros Lançamentos Caixa Input Projeto de Investimento Exportação de Investimento Input Carga de Depreciação Input faturamento entre Empresas DRE Input Cálculos de Pessoal Input Dados de Pessoal Manter Parametrização de DRE Fluxo de Caixa	To assess implemented improvements and identify new process and templates improvement opportunities focused on the adequacy to changes in strategic objectives.	Process and templates improved.	3
20 – 24	January to July 2012	POAB_Fase_2_Incremento_1 POAB_Fase_2_Incremento_2 POAB_Fase_2_Incremento_3 POAB_Fase_2_Incremento_4	To pilot incremental improvements (process and templates) and innovations (adaptation to small sized projects, code inspection, peer review of technical specification).	Quality and productivity improvement tendency. Process and templates adjusted to new objectives.	3 - > 4
25 - 26	October 2012	SCEBR-1754 SCEBR-1760 SCEBR-3024 SCEBR-4947 SCEBR-4949	To analyze subprocesses performance aiming to achieve subprocesses stabilization.	Quality and productivity improvement. Process and templates adjusted to new objectives.	4 - > 5
27 (Actual)	December 2012	SCEBR_1621 SCEBR_6811	To verify quality and productivity improvement. To pilot innovative improvement (peer programming and test coverage analysis).	Process analysis confirmed significant quality and productivity improvement were achieved in the projects.	5

In order to achieve quality and performance objectives established by Synapsis Brazil, first we conducted an analysis of the organizational process baselines that were adherent to CMMI level 3 practices adopted in the projects in the period of May to December 2011 (baselines 14 to 19 of Table 1). The analyses demonstrated that the quality and performance objectives were not being achieved and that several process improvements were necessary to achieve the goal. One of the main improvements conducted were the process optimization. This optimization was executed by automating process activities, and by eliminating redundant standard process activities that do not aggregate direct value to the software product. As a result, we obtained an optimized process with a minimum set of activities easy to understand, plan, monitor, measure and execute. After implementing the process optimization, 77% of the process activities were eliminated (104 out of 135 activities) without losing adherence to CMMI level 3 practices. This reduction were critical to allow the adoption of CMMI level 5 practices not only in small sized evolutionary projects, but also in large software development projects.

In the period of January to July 2012, were published baselines with not only the purpose of piloting incremental improvements in processes and templates, but also aiming to implement innovative improvements such as adaptation of the standard process to small sized evolutionary projects (less than 200 man-hour), and adoption of code inspection and peer review of technical specifications (baselines 20 to 24 of Table 1). The process adaptation strategy to small projects, and the adoption of practices to increase the quality of intermediate software development products (software architecture, test plans, requirement and technical specifications etc.) were considered critical success factors, because help to achieve the organizational quality and performance objectives (assertiveness of estimated productivity and reduction of defects detected by the client). Measurement analysis demonstrated that small projects were easier to be estimated, monitored and controlled, since corrective actions to manage deviations from project plans have more effect on the project than on large projects. Moreover, the anticipation of defects detection helped to improve product quality, and consequently, to reduce the amount of defects detected by the client in the acceptance testes. As a result of the pilot of these improvements, we could observe a tendency in the improvement of quality and productivity.

The next step was the stabilization of the processes by conducting performance analysis of selected subprocesses. The following subprocesses were selected for statistical process control since they are in the critical path of the projects: Specify product requirements, Elaborate technical specification and effort estimates, Code and integrate units, Conduct product tests and Conduct product acceptance tests. Improvements in processes and templates were adopted in new projects (baselines 25 and 26 of Table 1). The specialized knowledge of the consultancy team in high maturity practices, mainly knowledge about statistical process control were considered a critical success factor to conduct performance analysis and to evolve the process baselines. The knowledge acquired by taking specialized courses in the field helped to understand potential problems that affect process performance, and to suggest incremental and innovative improvements that could help to achieve organizational quality and performance objectives. Nevertheless, performance analysis of baseline data demonstrated that although the performance objective were achieved, it was not possible to achieve the quality objective, since many defects continued to be detected by the client in the acceptance tests.

The final step was to pilot innovative improvements like the adoption of peer programming and test coverage analysis in the projects (baseline 27). The performance analysis of the processes executed in the pilot projects confirmed the improvement in the quality and process performance. Consequently, we could successfully achieve the objectives established by the organization. The easy of process change acceptance and the tool support for statistical process control were considered critical success factors in this phase. During the conduction of the pilots, many improvements were implemented in the projects, including during projects execution, requiring some rework by the project managers and the development team. All project members were open to process changes, helping the organization to achieve its objectives. The adopted tools for statistical process control such as MiniTab™ and Oracle Crystall Ball™ reduced considerably the effort to perform data analysis; helping to rapidly verify if subprocesses attributes relations were statistically relevant.

DISCUSSION AND EVALUATION

In the last section, the knowledge about statistical process control, and the use of support tools to statistical process control were considered critical success factors in the implementation of high maturity practices in Synapsis Brazil. In order to manage those factors effectively, the process and templates of Synapsis Brazil were adjusted so that the application of statistical process control knowledge and techniques could be easily adopted by project managers during quantitative management of small sized evolutionary projects. This section presents some of the results obtained with the implementation of these improvements.

The lack of an adequate measurement repository in the market to support statistical process control was a problem for the implementation of statistical software process control. In order to minimize this problem, Synapsis Brazil implemented a specific measurement repository to support adoption of statistical software process control techniques. The adopted strategy in the implementation of the repository and in loading the necessary data was the following: (i) Defect information originated in Excel spreadsheets in which we implemented specific entries to facilitate collection of measures and to load data in the measurement repository; (ii) Information of estimated man-power effort, activities, process, size, complexity, productivity are loaded in the planning and monitoring tool used by project managers from size and effort data extracted of the estimation tool; (iii) Information of actual effort are loaded in the planning and monitoring tool used by project managers from data extracted of the Jira™ workflow tool; and (iv) Information about activities executed, estimated man-power effort, actual effort, activities schedule, productivity, product size, team, profile, technology etc., are loaded from measurement data extracted from the planning and monitoring tool. The processing of data stored in the repository, and two tools supported the visualization and generation of control graphics: Oracle Crystal Ball™ e Minitab™.

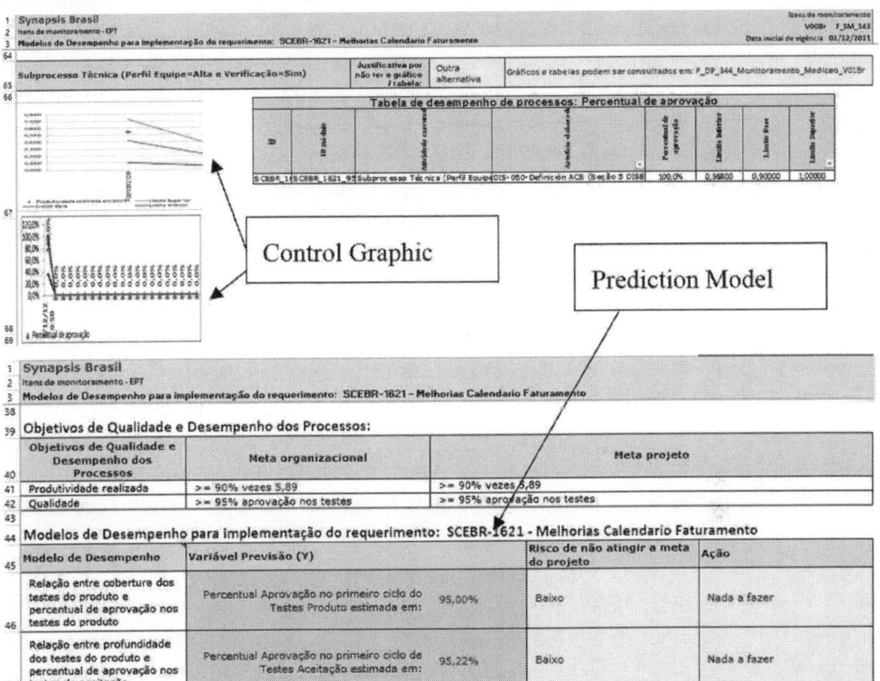

Figure 1. Example of a control graphic and prediction model to support software project quantitative management.

The lack of adequate tools to support quantitative management motivated the elaboration of a spreadsheet to support project managers in storing baseline and prediction models data. Figure 1 presents an example of the support tool to quantitative management. The tool presented in Figure 1 supports two statistical process control mechanisms. The top of the figure presents a control graphic automatically generated from project data. The bottom part of Figure 1 presents the probability analysis of not achieving the project quality and performance objectives based on the prediction models developed. By using the spreadsheet, statistical knowledge becomes transparent, thus facilitating its use and diminishing resistance by the project team.

It was observed that one of the most complex activities to be executed involved the selection of which subprocesses attributes should be controlled to address the established quality and performance objectives. The analysis of existence measurement data in the repository indicated that some of the characteristics of the subprocesses have greater influence in achieving the quality and performance indexes established by the organization. The combination of information about the characteristics of the subprocesses and other project information, such as characteristics of the team allocated in the project, allowed the project managers to verify the probability of achieving the quality and performance objectives every time the data were updated, thus providing the means to manage action plans to minimize associated risks of not achieving the objectives. This mechanism of subprocesses and its attributes selection not only simplified the implementation of a complex high maturity practice, but also facilitated the understanding of the process selection concept by the project managers.

Synapsis Brazil process group identified that the Minitab™ tool to support data regression analysis helps not only in the early stages of performance analysis, for instance, to verify correlation of subprocesses attributes selected for statistical process control, but also supports the calculation of prediction models adopted in the quantitative management of Synapsis Brazil projects. For instance, the following performance model was developed by the application of regression analysis technique supported by MiniTab™: *Percentage of Approved = 0,818 + 0,0717 * Amount of Criteria by Function Point*. This model indicates that exists a positive correlation among the amount of criteria by function point in the product tests and the percentage of test cases approved in the product acceptance tests by the client. One of the problems of application of regression analysis is the identification of outliers, i.e., specific data points that do not represent a common behavior in the project. MiniTab™ helped to automatically identify the outliers, thus facilitating the elimination of anomalies in the data that hindered the development of trustful models.

We also noticed that the sensitivity analysis technique of Monte Carlo simulation using Oracle Crystal Ball™ tool helped to identify critical subprocesses that affected the achievement of quality and process performance objectives of Synapsis Brazil, such as Product Tests subprocesses. After the implementation of incremental and innovative improvements in those subprocesses, we could verify a significant change in the processes performance, helping the organization to achieve its objectives. One of the problems in the application of Monte Carlo simulation was the lack of past project data necessary for conduction of simulations. As the understanding of critical subprocesses and its relationships increase, new measures are necessary for performing the analysis. Nevertheless, many times these measures are not found in the repository. This problem forced the elimination of several projects in the early stages of baselines data analysis due to lack of data that could be used in the simulation analysis.

The implementation of software process improvement is a complex activity that requires significant amount of human and financial resources. Moreover, the return of investment is observed in a long-term period. Nevertheless, software process improvement is likely to pay-off when the improvements are adopted across the organization and software development maturity is consistently observed during projects execution. We could observe some significant results after successfully conducting software process improvement initiatives in Synapsis Brazil. Table 2 presents some of the direct results in defect density, productivity and schedule observed after implement CMMI maturity levels 2, 3 and 5.

Table 2. Defect density, productivity and schedule improvement results in Synapsis Brazil after implementing CMMI levels 2, 3 and 5

	Defect density		Productivity		Schedule	
	Defects/ function point	% reduction	Man-hours/ function Point	Improvement	% projects delayed	Improvement
CMMI 2	0,65	-	13,5	-	45%	-
CMMI 3	0,38	42%	10,9	19%	20%	56%
CMMI 5	0,32	16%	9,3	15%	4%	80%

Table 2 shows that there was a significant reduction of defect density from level 2 to level 3 (42%). This was considered a good result since the focus of CMMI level 3 process areas is product engineering and verification and validation practices. Although the reduction of defect density from CMMI level 3 to level 5 were smaller (16%), Synapsis Brazil higher management was satisfied with the result since it was possible to achieve the business objectives concerning product quality.

With respect to productivity, we could observe a continuous improvement from level 2 to 3 (19%) and from level 3 to 5 (15%) in Table 2. One of the causes of this result was the process optimization performed during CMMI level 3 and 5 implementation thus reducing the overall effort to develop software.

The best results were observed in the percentage of delayed projects in Table 2. This percentage dropped from 45% of delayed projects to 20% after implementing CMMI level 3 and continued to drop to 4% of delayed projects after implementing CMMI level 5. The reduction of project size can be one of the causes of this significant improvement. Moreover, the implementation of quantitative project management practices and statistical process control techniques helped to prevent future process performance and to take more effective corrective actions during projects execution.

The Synapsis Brazil process group observed as a lesson learned the fact that the process implementation aiming high maturity levels do not represent greater effort related to CMMI level 3 implementation. The project team is not significantly affected by the implementation, since the new process areas to be implemented in the standard processes do not affect directly the team. The measurement group is the most affected group by high maturity implementation since they have to assist project managers, some members of the project team and the higher management with control information so that improvement actions can be planned and executed in time to correct deviations before concluding projects activities. Another issue that the process group considered as a differential was the facility that the project manager had to initiative the project planning concerning subprocesses definition that compose the adapted process to the project. The activity of subprocesses selection is fundamental to assess achievement of quality and performance objectives with the composition of the project process. This activity was facilitated by the automated support in the selection of subprocesses based on project characteristics and organizational performance baselines.

Project managers observed that the available tools and process and quality groups were the main structure to support quantitative management, providing the mechanisms for planning and controlling projects activities, i.e., those groups not only supported project managers in the effective planning, but also supported the project team to solve problems during projects execution. Risks planning and monitoring helped to increase the understanding of factors that may jeopardize projects execution concerning achievement of quality and performance objectives. A continuous monitoring minimizes the occurrence of problems during projects execution, since it is fundamental to be aware of the situation of projects products, team and client commitments and accomplishment of established milestones and budget. The project managers also noted that the process efficacy is important but the team involvement and commitment to work have also great influence in projects success. People continue to be the essential part of the process and project managers should always keep that in mind. In this work, the great differential was the use of statistical data. Project managers noted that using these data in an adequate manner increases the chances of predicting future process performance aiming to facilitate achievement of project objectives.

The consultancy observed that the lack of data jeopardizes the assessment of process performance. A way to overcome this problem is to focus the analysis in evolutionary projects of small size, because they allow rapid data generation. The development of initial prediction models with few independent variables (one to two at most), also helps to obtain more trustful models when few data is available.

CONCLUSION

This paper presented the experience of Synapsis Brazil in the evolution of its processes to achieve high maturity. We presented the main critical success factors, such as easy to accept changes by project teams, support and commitment of higher management, adequate supporting tools for statistical process control, incremental and innovative improvements and the specialized knowledge of the consultancy in statistical process control. The paper also discussed the results of adopting statistical process control in small sized evolutionary projects. Lessons learned by the teams were also discussed.

Considering that the purpose of CMMI level 5 is continuous optimization, the next steps for Synapsis Brazil process improvement is to analyze performance of the actual baseline aiming to identify the subprocesses that affect the achievement of quality process performance objectives reviewed by the higher management aiming a more competitive position both in the national and international software market.

COMPETING INTERESTS

The authors declare that they have no competing interests.

AUTHORS' CONTRIBUTION

MM identified common problems of similar cases, elicited some critical barriers for conducting the reported experience and drafted the manuscript. CS documented the case results and lessons learned. Both authors read and approved the final manuscript.

REFERENCES

1. Florac WA, Carleton AD, Barnard JR: *Statistical process control: analyzing a space shuttle onboard software process. IEEE Software 17:97–106. n 4, July/August*. CA, USA: IEEE Computer Society Press Los Alamitos; 2000.
2. Kitchenham B, Kutay C, Jeffery R, Connaughton C: *Lessons learnt from the analysis of large-scale corporate databases. ICSE '06 Proceedings*

of the 28th international conference on Software engineering. NY, USA: ACM New York; 2006:439–444.

3. Luo W, Wang M, Zhou B, Liu P: *Research on CMMI-based project management environment, 2008 International Conference on Wireless Communications, Networking and Mobile Computing*. Dalian, China: WiCOM 2008; 2008.

4. Montoni MA, Rocha AR: Using grounded theory to acquire knowledge about critical success factors for conducting software process improvement implementation initiatives. *Int J Knowl Manag (IJKM)* 2011, 7(3):43–60.

5. Sargut KU, Demirors O: Utilization of statistical process control (SPC) in emergent software organizations: pitfalls and suggestions. Software Quality Journal. *Kluwer Academic Publishers* 2006, 14(2):135–157.

6. SEI – Software Engineering Institute: CMMI for SCAMPI class a appraisal results 2012 Mid-year. 2012.http://cmmiinstitute.com/assets/presentations/2012%20SepCMMI.pdf Update, , accessed in April/2013

7. SOFTEX: Total number of organizations with MPS Assessment (valid or not valid) (in Portuguese). 2013.http://www.softex.br/mpsbr/_avaliacoes/avaliacoes_mpsbr_total.pdf , accessed in April/2013

8. Travassos GH, Kalinowski M: *iMPS 2011 Performance Results of Organizations that Adopted the MPS Model from 2008 to 2011 (in Portuguese)*. Campinas/SP: SOFTEX; 2012:36. , accessed in April/2013 (2012) http://www.softex.br/wp-content/uploads/2013/08/iMPS-2011-Resultados-de-Desempenho-das-Empresas-que-Adotaram-o-Modelo-MPS-de-2008-a-2011.pdf , accessed in April/2013 (2012)

9. Weller E, Card D, Curtis B, Raczynski B: *Point/counterpoint applying SPC to software development: where and Why, software. IEEE Software 25(3):48–51. May-June*. CA, USA: IEEE Computer Society Press Los Alamitos; 2008.

Index